GENE TARGETING

GENE TARGETING

John M. Sedivy

Department of Molecular Biophysics and Biochemistry
Yale University School of Medicine

Alexandra L. Joyner

Samuel Lunenfeld Research Institute
Division of Molecular and Developmental Biology
Mount Sinai Hospital

Department of Molecular and Medical Genetics
University of Toronto

W. H. Freeman and Company
New York

Figure on p. 4 adapted from Arthur Kornberg, *DNA Replication,* First Edition (New York: W. H. Freeman, 1980), Fig. 1-8.

Figures on pp. 10, 26, and 27 adapted from Bruce Alberts et al., *Molecular Biology of the Cell,* First Edition (New York: Garland, 1983), Figs. 3-16, 8-5, and 14-15, respectively.

Library of Congress Cataloging-in-Publication Data

Sedivy, John M.
Gene targeting / John M. Sedivy, Alexandra L. Joyner.
p. cm.
Includes bibliographical references and index.
ISBN 0-7167-7013-X
1. Genetic engineering. I. Joyner, Alexandra L. II. Title.
QH442. S43 1992
575.1'0724—dc20 91-44340
CIP

Printed in the United States of America

2 3 4 5 6 7 8 9 0 HA 9 9 8 7 6 5 4 3 2

To my teachers:
Dan Fraenkel, Phil Sharp, and Alan Weiner

—John Sedivy

To Dan and Rowena
who at times played second fiddle to
my obsession with gene targeting

—Alexandra Joyner

CONTENTS

PREFACE

DNA is a marvelous molecule. It is the blueprint and the archive, handed down virtually unchanged from generation to generation, that encodes to a large degree the information that specifies our physical form. The encoded information is continually accessed, read out, and executed. Many programs run concomitantly, some continuously, others only during specific times, for example, during development, or in response to various environmental stimuli. All living creatures refer to their DNA for instructions in a constant and dynamic give and take. The contribution of genetic versus environmental influences in the shaping of intelligence, complex behavior, and personality has not been and may never be precisely determined.

Gene targeting holds out the possibility of being able to tap into and change the information contained in DNA. Modifying DNA is not a novelty: selective breeding, as practiced in the cultivation of agricultural crops and animal husbandry, is simply an indirect and rather inefficient way of going about it. In contrast, gene targeting is a direct intervention at the molecular level, precise and specific.

In the study of molecular biology, we are faced with a peculiar quandary: we seem to know so much, yet often we can understand so little. To a certain extent, this is inherent in the experimental approach; molecular biology is, after all, a study from the bottom up. Viewed this way, it may thus not come as so great a surprise that we now understand the alphabet and even know quite a few words, but the whole story still escapes us.

We set out to write this book with one main purpose in mind: to explain, in as simple terms as possible, what gene targeting is. Where did it begin and what is its history? What are its underlying principles? What makes it possible? How is it done? What is it being applied to? Our goal was to provide answers to these and related questions without going into the details of all the underlying experimental evidence. At the same time we tried to express a feeling for the conceptual beauty of molecular biology and the real excitement of scientific discovery.

Achnowledgments

John Sedivy would like to thank Charles Radding, David Gonda, and Mike Liskay for very helpful discussions as he was writing his chapters. Alexandra Joyner would like to thank Janet Rossant for her insights into mouse embryogenesis and for encouragement to persist until gene targeting was successful. She is also grateful to the members of the laboratory for their input into her way of thinking about homologous recombination. Finally, she wishes to thank Bob Joyner and Wolfgang Wurst for their comments on the manuscript and Sherry Mackey for promptly typing the many drafts.

1

DNA

Hundreds of chapters like this one have been written. Nevertheless, for meaningful plot development, the leading character must be properly introduced.

The sole function of DNA is information storage. Once executed, this information specifies a three-dimensional form of an organism in space as well as in time, yet the machine language is linear and consists of an alphabet of four symbols and no punctuation marks. The analogy with computer design is helpful: An intricate picture drawn with a sophisticated graphics program is stored in computer memory as a linear sequence of two symbols, 0 and 1. To re-create the picture one needs three components: the machine language record, a program capable of its execution, and the hardware to accomplish the task.

Structure of DNA

As one may have guessed from the computer analogy, DNA is a linear molecule consisting of numerous single, or monomer, units joined in a beads-on-a-string fashion. As seen in Figure 1.1, the monomer beads, called **nucleotides,** have three chemical components or groups: a **phosphate,** a **carbohydrate** (**deoxyribose**), and a **base.** The term base in this context simply denotes that the group is, chemically speaking, basic (alkaline), as opposed to acidic. The bases come in four flavors: **adenine, guanine, cytosine,** and **thymine.** They constitute, in fact, the four-

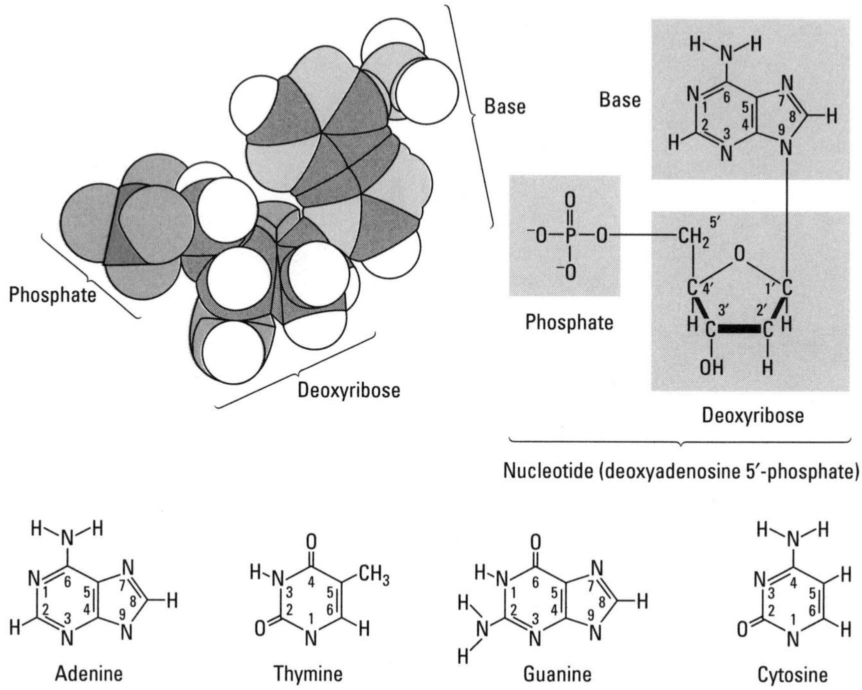

FIGURE 1.1 Building blocks of DNA. The chemical structure of the nucleotides is shown. Only five kinds of atoms go into making a nucleotide: carbon, oxygen, nitrogen, hydrogen, and phosphorus. The atoms are joined together by covalent chemical bonds, represented by lines in the drawings. Covalent bonds are very strong; as a result, the structure of the nucleotides is quite stable. An assembly of atoms joined by covalent bonds into a stable structure is often referred to as a molecule. At top left, a nucleotide molecule is shown in a more lifelike fashion, called a space-filling model. All nucleotides are based on ring structures, either a six-member ring, called a pyrimidine, found in cytosine and thymine, or a fused five and six-member ring, called a purine, found in adenine and guanine. Thus, cytosine and thymine are both pyrimidines, and adenine and guanine are both purines.

symbol alphabet, namely the beads. The deoxyribose and phosphate groups are constant and merely form the string.

Both the bases and the carbohydrate are ring structures, and each position of the ring is numbered for orientation. The deoxyribose ring is numbered 1′ (one-prime) through 5′ (five-prime) to distinguish it from the numbering of the bases. In a conventional mononucleotide, the phosphate group is attached at the 5′ position of deoxyribose. When joined in a string, or **polynucleotide,** each phosphate is attached to the 3′ deoxyribose position of its neighbor (see Fig. 1.2).

It is important to note that nucleotides and thus resultant polynucleotides have a three-dimensional form as well as handedness. This means that mirror images are not identical, the same as, for example, this

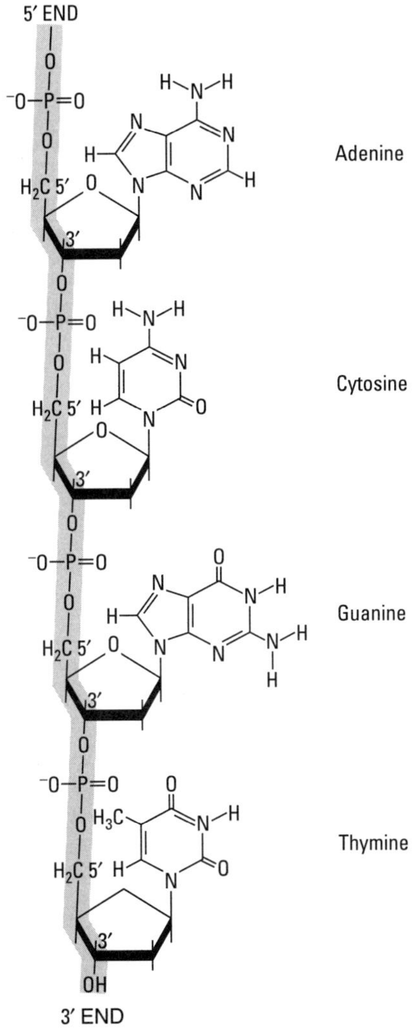

FIGURE 1.2 Structure of polynucleotides. The chemical structure of a short polynucleotide chain is shown. Each phosphate is covalently bonded to its two adjacent deoxyribose neighbors; thus, the four monomer nucleotide molecules are joined into a larger molecule. Such assemblies can get very large indeed and are often referred to as macromolecules.

sentence, which viewed in a mirror will appear backwards, or right to left. The information encoded in a polynucleotide can be written simply as the sequence of its bases, which are abbreviated A for adenine, T for thymine, G for guanine, and C for cytosine. Because of handedness, the orientation of the sequence, namely its 5′ and 3′ ends, must also be specified (see Fig. 1.2). For example, in three dimensions, the octanucleotide 5′-GTCCATTC-3′ is a different structure from 3′-GTCCATTC-5′. By convention, DNA sequences are always written in a 5′→3′ direction.

The property of DNA that allows it to function as the molecular archive is its ability to be duplicated. Each time a cell divides, both daughters must receive a complete copy of the blueprint of life. This is possible because two polynucleotide chains can form a side-by-side, double-stranded structure, the bases contacting each other in the middle and the deoxyribose–phosphate backbones running on either side (see Fig. 1.3). The key to this structure is that not all bases can pair with each other: A will only pair with T, and G will only pair with C. In addition, the two single chains can associate only in an antiparallel fashion, one chain running in the 5′→3′ direction and the other in the 3′→5′ direction.

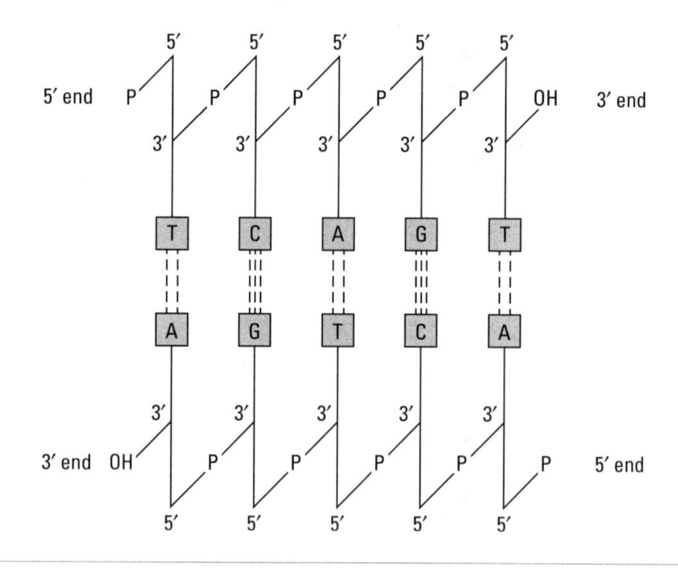

FIGURE 1.3 Double-stranded DNA. Two-dimensional representation of a short stretch of double-stranded DNA, showing the antiparallel orientation of the two polynucleotide chains. Two hydrogen bonds form between A–T base pairs, while three bonds form between G–C base pairs. Thus, G–C base pairs are stronger, and as a result, GC-rich double-stranded DNA is more stable and harder to separate into its two polynucleotide chains.

For example, the octanucleotide GTCCATTC, when converted to its double-stranded form, would appear as

5′-GTCCATTC-3′

3′-CAGGTAAG-5′

The bottom strand, GAATGGAC (written in the conventional 5′→3′ fashion), is referred to as the **complement** of the original octanucleotide (GTCCATTC). In fact, each contains all the information required for its exact duplication, inherent in the two pairing rules: (1) specificity (A with T, G with C), and (2) directionality (5′→3′ with 3′→5′).

The interactions between base pairs in double-stranded DNA take the form of weak chemical bonds, called **hydrogen bonds.** These bonds are much weaker than the **covalent bonds** holding together the groups within a single polynucleotide chain. Added up over long stretches, however, the weak hydrogen bonds result in a very stable double-stranded structure. It is the hydrogen bonds, in fact, that dictate the pairing rules; good bonding occurs only between the correct bases and only in a proper three-dimensional orientation. The specificity of the hydrogen bonds is in turn dictated by the chemical composition of the bases.

The picture of DNA shown in Figure 1.3 is a schematic two-dimensional representation. Double-stranded DNA, however, in three dimensions exists as a **double helix** (see Fig. 1.4). Picture the flat representation as a ribbon, the backbone of each strand forming one of the edges; now twist the ribbon around a cylinder to form a helix, giving a barber-pole appearance. The base pairs are oriented perpendicular to the longitudinal axis of the helix, filling the inside of the cylinder, while the deoxyribose–phosphate backbones face outside.

In most cells, DNA is found in very long double-helical chains, millions of base pairs long. At each cell division, the two polynucleotide strands are separated, and a new complementary strand is synthesized along each of the parental strands (see Fig. 1.5). The result is two identical double-stranded molecules, each composed of one parental and one newly synthesized strand. This mode of replication is often referred to as **semiconservative.** Each daughter cell then inherits one of the double–stranded molecules.

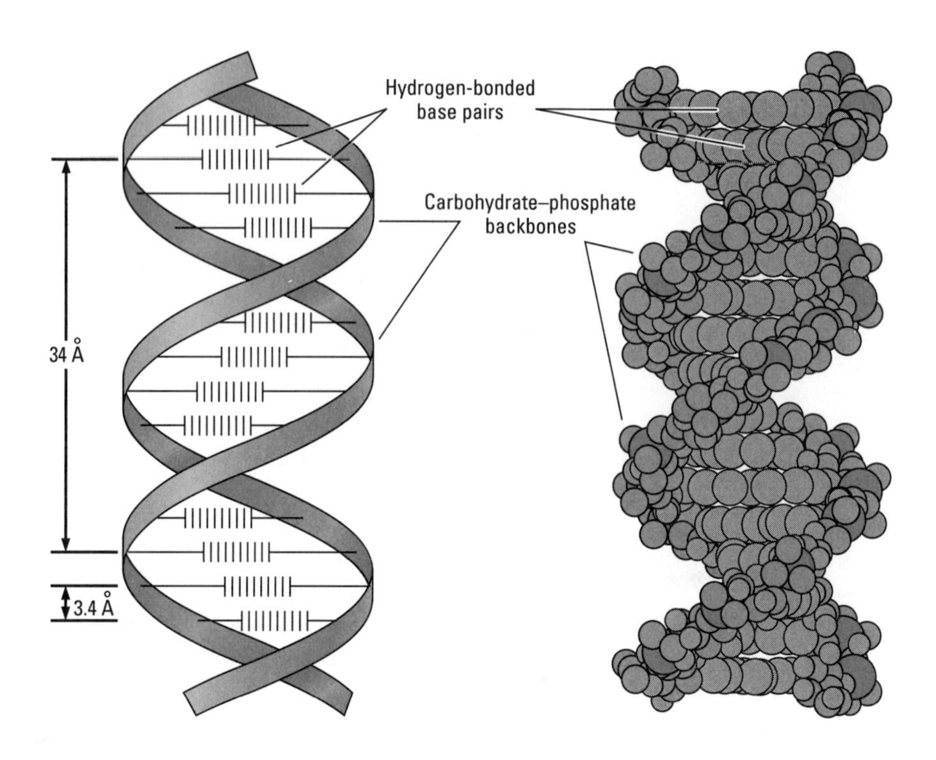

FIGURE 1.4 DNA double helix. Three-dimensional structure of the DNA double helix and the corresponding space-filling model. One complete turn of the helix is 34 Å long when measured along the longitudinal axis (often referred to as the pitch). Å (Ångstrom), one ten-billionth (10^{-10}) of a meter, is a unit often used on the molecular distance scale. The center-to-center distance between neighboring base pairs is 3.4 Å. The base pairs are flat and adopt a stacked configuration at the center of the helix, not unlike dominoes piled one on top of each other. Two conventions can be used to describe the configuration of a helix: clockwise versus counterclockwise, or right-handed versus left-handed. For example, to insert a corkscrew or screw in a lightbulb, a clockwise motion is required. These are thus examples of clockwise, or right-handed, helices. The DNA double helix is also a right-handed helix. The form of DNA shown here is referred to as the B form. This is the form found most commonly in living cells. Under special conditions, however, DNA can adopt several different conformations.

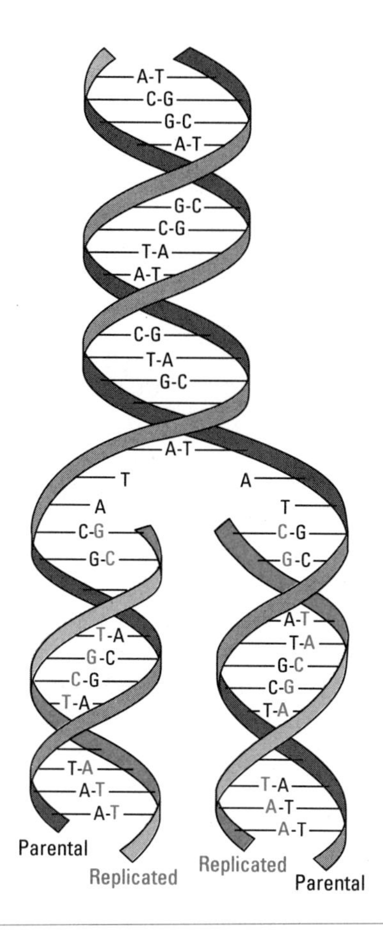

FIGURE 1.5 Replication of DNA. The parental double helix, pictured in grey, becomes unwound into its two polynucleotide chains. New complementary polynucleotide chains are laid down along the separated chains according to the two pairing rules, specificity and directionality. The synthesis of the new DNA chains is catalyzed by a complex enzymatic machinery. The key enzyme, DNA polymerase, uses the parental strand as a template, moving along it in stepwise fashion and polymerizing nucleotide monomers into a complementary polynucleotide. Multiple enzymes participate in the reaction in auxiliary roles. For example, the mononucleotides need to be first converted to triphosphates, a high-energy activated form. Replication of DNA is always initiated at sites called **origins,** which contain specific DNA sequences that are recognized by some of the auxiliary enzymes working along with DNA polymerase. Once replication has been initiated it spreads outwards from the origin into the surrounding regions of DNA.

Function of DNA

Although DNA has been referred to as the archive, it is not an heirloom handed down from generation to generation, spending most of its time on a shelf gathering dust. From conception to death, the information encoded in DNA is being continually accessed, read out, and executed. Using the computer analogy, many programs run concomitantly, some continuously, others only during specific times, for example during development or in response to various environmental stimuli. All living creatures constantly refer to their DNA for instructions, in a constant and dynamic give-and-take. The information, however, flows in one direction only. In other words, the "software" is used in a strictly "read-only" mode, since environmental stimuli and experiences never become directly encoded in DNA.

All life functions can ultimately be broken down into chemical reactions. Living organisms must integrate a vast number of chemical reactions, and the regulation must be precise in both space and time. This capacity for control has allowed the evolution of a rich variety of life on this planet as well as an exquisite complexity of many biological functions. Although many chemical reactions are spontaneous, most biologically important reactions must be **catalyzed** to occur at an appreciable rate. A **catalyst** increases the rate of a reaction without being itself consumed or altered. Most biological catalysts are **enzymes,** although other molecules can in certain circumstances have catalytic activity.

Enzymes are the most important part of the hardware of the cell. Chemically speaking, enzymes are made of **protein.** Proteins are composed of monomeric subunits called **amino acids** and take the form of long unbranched chains. Among all the life forms on this planet, only 20 amino acids are commonly used. Conceptually, this is similar to the way DNA is made of monomeric nucleotides, except that proteins are relatively much shorter, on the order of hundreds rather than millions of monomers.

This similarity is not coincidental. When information encoded in DNA is "read out," it is first **transcribed** into **RNA** (see Fig. 1.6). RNA is a polynucleotide similar to DNA; the only difference is that a slightly different carbohydrate is used in the backbone, **ribose** instead of deoxyribose, and that the base thymine is replaced with the base **uracil.** These subtle differences, however, have profound effects on the chemical reactivity of RNA. In most cases, RNA is a copy of the sequence information in a particular stretch of DNA. RNA molecules can function in three ways: information transfer, catalysis, and structure. In its catalytic and

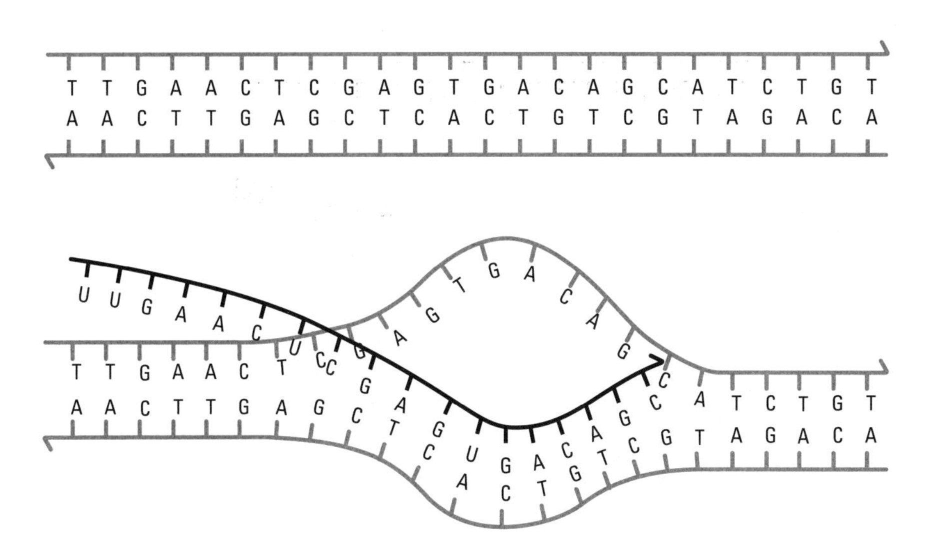

FIGURE 1.6 Transcription of DNA into RNA. The DNA double helix becomes unwound, and a single complementary chain is laid down along the template strand. There are several notable differences between transcription and replication. First, the newly synthesized polynucleotide is RNA, not DNA. Second, only one of the DNA strands is copied. Third, the unwinding of the DNA double helix is only local. The newly synthesized RNA chain is dissociated from the template DNA strand, which then resumes its usual pairing with its DNA partner. The whole process is accomplished by a single complex enzyme, RNA polymerase.

structural modes, RNA molecules function in concert with specialized proteins as molecular factories that carry out specific generic tasks inside the cell.

In its information transfer mode, RNA, called messenger, or **mRNA,** is **translated** into protein (see Fig. 1.7).* Each successive three nucleotides encode one amino acid. This is, in fact, the **genetic code:** a triplet of nucleotides forms a symbol for a particular amino acid. Since there can be a total of 64 distinct triplets of four nucleotides ($4 \times 4 \times 4 = 64$), some amino acids are encoded by more than one triplet. A triplet of nucleotides that specifies an amino acid is commonly referred to as a **codon.** The

* Common words of everyday language often assume very specific meanings in the jargon of molecular biology. Two such important cases are emphasized here. *Transcription* always means the copying of DNA to RNA, while *translation* always refers to the copying of RNA to protein.

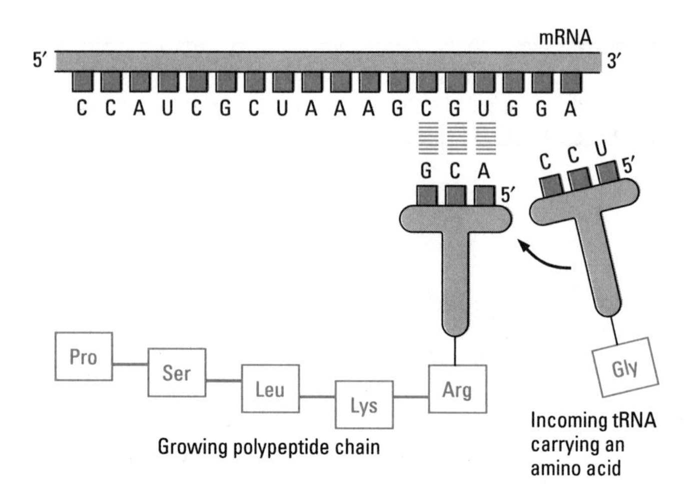

FIGURE 1.7 Translation of mRNA into protein. The mRNA polynucleotide forms a scaffold along which amino acids can be aligned in a specific order and joined together into a polypeptide chain. The recognition between a nucleotide triplet in the mRNA, a codon, and its cognate amino acid is mediated by specialized RNA adaptors, the transfer RNAs (tRNA). Translation is a very complex process catalyzed by a large macromolecular complex, the ribosome. A ribosome is composed of three distinct polynucleotide chains of RNA (rRNA) and over fifty distinct proteins. Additional enzymes act as auxiliary components of the translation machinery, for example, to attach amino acids to their specific tRNAs.

genetic code contains 61 codons; the remaining three triplets are utilized as signals that instruct the protein translation machinery to terminate synthesis.

Most of the information in DNA encodes proteins. Proteins are the workhorses of the cell. In their catalytic roles, proteins as enzymes accomplish much of the work of the cell, such as all metabolic activity, replication of DNA itself, transcription of DNA into mRNA, and so forth. For example, when yeast cells ferment sugar into alcohol, the process takes place as a stepwise series of chemical reactions, each catalyzed by a distinct enzyme. In their structural roles, proteins form much of the bricks and mortar of our bodies. For example, hair is mostly composed of a specialized structural protein called keratin, while connective tissue contains another protein called collagen.

The unit of DNA that encodes the information for one enzyme is called a **gene.** The "one gene–one enzyme" rule is one of the cornerstones of molecular biology. A bacterium, such as *Escherichia coli,* contains approximately three thousand genes, while a mammal probably contains on the order of fifty thousand genes. A gene contains several functional units (see Fig. 1.8). The body of the gene is comprised of the **coding region** that contains the information for synthesis of the gene product, for example, a polypeptide chain. In front of the gene is a region referred to as a **promoter,** which serves as a recognition point for RNA polymerase and

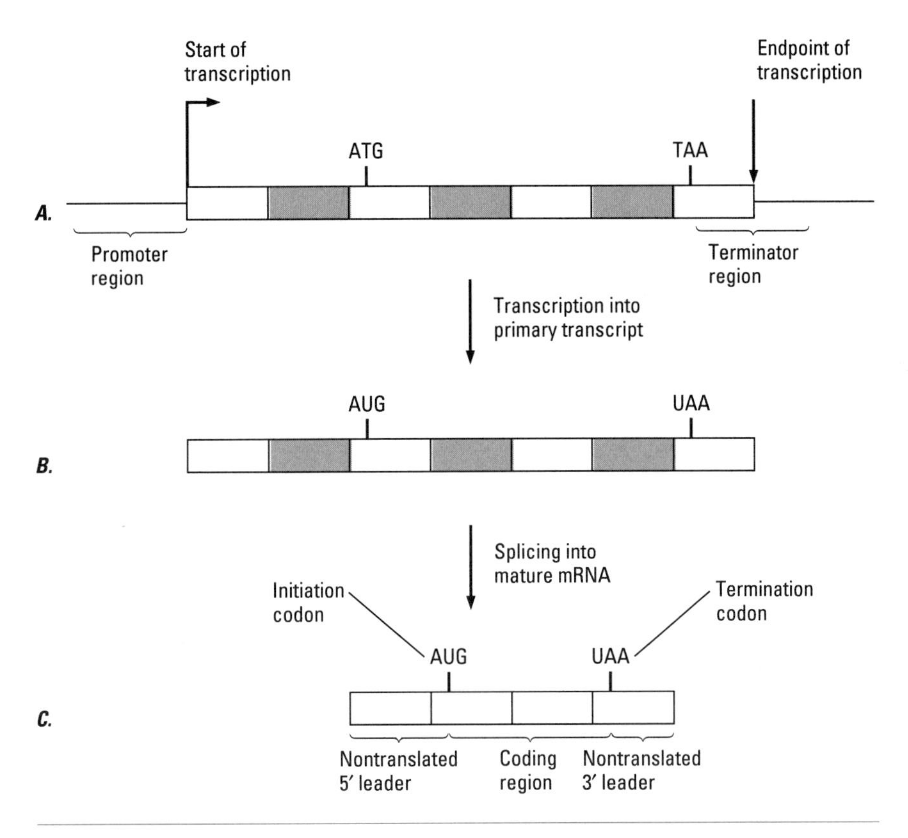

FIGURE 1.8 The structure of a gene. A typical gene in higher organisms contains introns and exons. Exons are shown as white boxes, introns as green boxes, and nontranscribed surrounding chromosomal sequences as black lines. A gene is preceded by a promoter region and followed by a terminator region. The start of transcription defines the beginning of the first exon. A gene may contain many exons and introns, and introns may be many thousands of base pairs long. Genes come in a wide range of sizes, anywhere from less than a thousand to over a million base pairs. Primary transcripts can thus be very long and comprised primarily of introns. Splicing must be precise down to the single nucleotide level to ensure that the joined exons contain the correct reading frame through the entire coding region. The final mature mRNA often contains nontranslated regions, called leaders, on its 5′ and 3′ ends. Translation is almost always initiated at an AUG codon, which is shown here at the beginning of exon 2. A terminator codon, UAA, is shown in exon 4.

auxiliary factors. Successful recognition of the promoter results in the initiation of transcription of the coding region into mRNA. Behind the gene is another specific recognition sequence, called a **terminator,** that signals the RNA polymerase to stop transcription. Genes in bacteria have this simple structure, but genes of higher organisms often contain interruptions in the coding region called **introns.** Introns are noncoding portions of the gene, and although initially transcribed into mRNA, they are subsequently removed by a process usually referred to as **splicing.** The coding regions, called **exons,** are thus joined together during the splicing reaction to generate an uninterrupted coding region that can be successfully translated.

A gene is said to be **active** or **on** when it is being transcribed into mRNA. The "programs" that determine, for example, embryonic development in fact specify gene activity; in other words, what genes, when in time, and where in the body, are on or off. The study of regulation of gene activity is a major and exciting field in modern molecular biology. A more detailed discussion of that topic is, however, beyond the scope of this book. This chapter has merely sought to establish a general understanding of a framework, an "operating system," within which the rest of the book will be confined.

2

EVOLUTION

It makes good sense that DNA should be handed down from generation to generation as unchanged as possible, since what good would be an archive that is constantly changing? On the other hand, if the DNA archive was absolutely constant and unchangeable, evolution would be impossible. This is, in fact, a crux of a dilemma: the preservation of the tried-and-true information in face of the need for adaptation to a constantly changing environment. Specific molecular mechanisms are responsible for maintaining a delicate balance in this tug-of-war of opposing forces.

Mutation Is the Driving Force of Evolution

Any change in the DNA sequence from what previously existed is called a **mutation.** The term mutation does not imply anything about the eventual effect on the organism, which could be beneficial, detrimental, or neutral. In terms of actual changes in the DNA sequence, mutations come in many varieties. One common type occurs by substitution of one nucleotide for another; another happens when a nucleotide goes missing or an extra nucleotide is picked up (see Fig. 2.1). These are referred to as **point mutations,** since they are very localized in the context of long DNA chains. Other types of mutations, often referred to as **rearrangements,** are more gross in nature: A DNA segment, often very large, may be removed (**deletion**), placed in

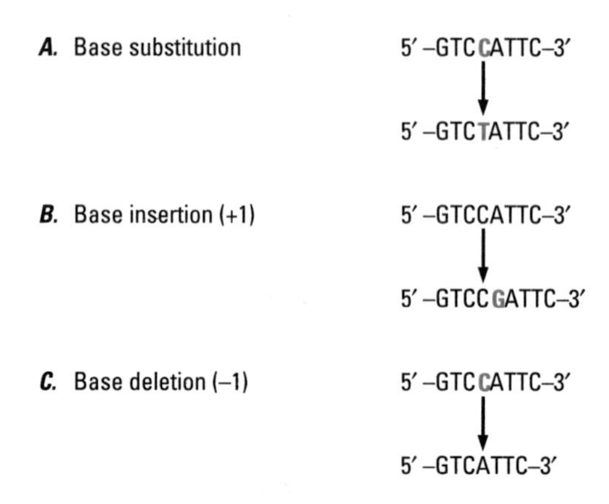

FIGURE 2.1 Point mutations of DNA. Point mutations were originally defined as lesions that mapped to a single point on a genetic map and could not be further resolved. In molecular terms, point mutations refer to lesions involving only one, or at most a few, base pairs of DNA. Three general types can be envisioned: (*A*) a substitution of one nucleotide with another, (*B*) a deletion of a nucleotide, and (*C*) an insertion of a nucleotide. When such mutations happen in regions of DNA that code for proteins, they are often described in terms of their effect on the encoded gene product. Base substitutions that change one codon to another are called missense mutations; changes from a codon to a termination signal are called nonsense mutations. Base deletions and insertions are often referred to as frameshift mutations.

another location (**insertion**), reciprocally exchanged with another segment (**translocation**), and so forth.

DNA is quite an inert molecule. Such chemical stability makes it ideally suited for its archival role. Its replication is accomplished by a host of specialized enzymes, loosely referred to as the replication machinery. The replication machinery is indeed extremely precise, committing only one mistake per approximately 10 million base pairs replicated. Most point mutations arise as errors in replication; many chemical agents can substantially increase the error frequency. The mechanisms by which rearrangements occur are more obscure; some are due to physical damage to DNA, which can be caused by, for example, X rays.

Without mutation there would be no evolution. For continued adaptation, multiple successive mutations must occur. Mutations leading to beneficial changes must be retained, and detrimental mutations must be discarded. Organisms carrying beneficial mutations must then undergo further rounds of the same process; in other words, mutations must be tried in various combinations. If evolution occurred only through mutation, however, it would be an inefficient process.

Sexual Reproduction Greatly Facilitates Evolution

The second major force of evolution is sexual reproduction, which literally allows organisms of the same species to compare notes on the information content of their DNA. Sex provides a mechanism for the rapid spread of beneficial mutations through populations, as well as the testing of the combinatorial effects of multiple mutations.

Most higher organisms contain more than one complete copy of their DNA archive—usually two. This makes good sense: One copy comes from the father and one from the mother. The entire sum of one's DNA is called the **genome;** if it contains two complete copies a genome is said to be **diploid.** When the time comes to reproduce, however, only one copy can be passed on. Otherwise, the number of copies (**ploidy**) would double with each generation. But which copy is to be passed on, the paternal or the maternal? If pieces of both could be combined into exactly one entire copy, or **haploid genome,** beneficial mutations in the maternal and paternal complements could be transmitted together to a single individual in the next generation. Such shuffling of genetic information is, in fact, achieved by two mechanisms: **independent reassortment** and **recombination.**

The haploid human genome contains 3 billion (3×10^9) base pairs of DNA, which is too large to exist as a single, continuous double helix. The genome is therefore divided into smaller parcels, called **chromosomes,** which are given simple numerical names. The largest human chromosome, number 1, contains approximately 260 million (2.6×10^8) base pairs of DNA, and the smallest, number 22, approximately 50 million (5×10^7) base pairs. A normal diploid human cell contains two copies of each chromosome 1 through 22 and two **sex chromosomes.** The sex chromosomes, called X and Y, are haploid in male cells, while female cells are diploid for the X and lack the Y chromosome. The two copies of a particular chromosome are said to be **homologous,** since they are virtually identical in sequence. The sequence identity of homologous chromosomes is very high, but not absolute. Since each copy was inherited from one parent, it contains the mutations that occurred in that individual, as well as some of the mutations from previous generations.

Mitosis and Meiosis

Distinct mechanisms have evolved at the cellular level to achieve, on the one hand, sexual reproduction of the species and, on the other, growth and development of an organism by asexual cell division. When cells in the body divide, each of the daughters is a carbon copy of the parent. In preparation for division, the DNA of the cell is duplicated (see Fig. 2.2); shortly afterwards, the two copies of the genome are partitioned and the

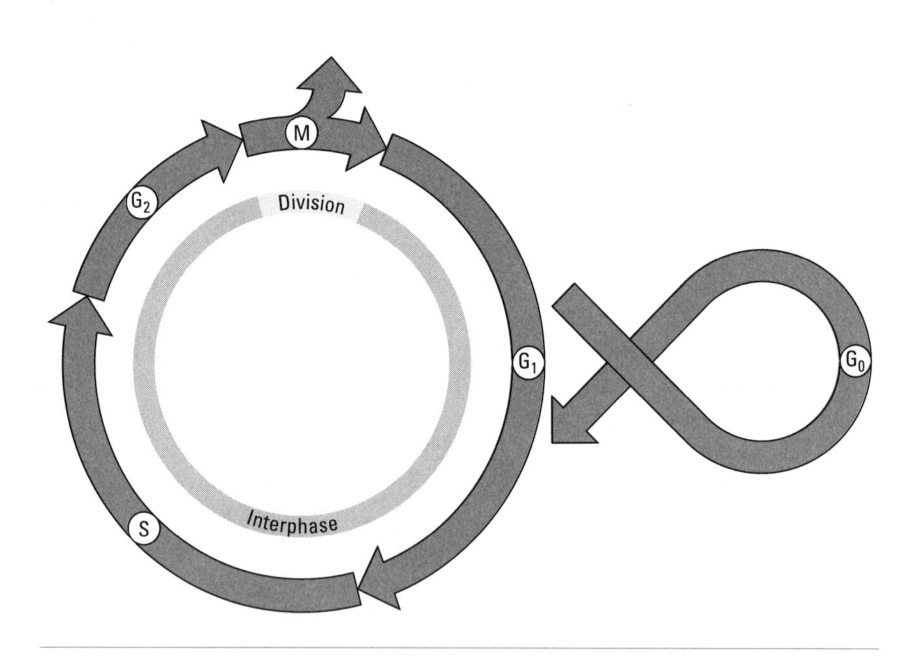

FIGURE 2.2 Cell cycle. Immediately after division, cells enter a period of growth, designated growth phase 1, or G1. Actively growing cells use this interval to prepare for the next round of division. Cells can also make a decision to withdraw from the cell cycle into a resting state, often referred to as the G0 phase. During the sythesis, or S, phase, the DNA is replicated. Another period of growth follows, designated G2, when final preparations for ensuing cellular division are made. Chromosomes are partitioned and daughter cells are formed during a brief period called mitosis (see Fig. 2.3). The ploidy of the cells in G2 is double that in G1, and mitosis in turn halves it.

two progeny cells become physically separated. This asexual process is called **mitosis** (see Fig. 2.3). Importantly, the ploidy of the genome remains unchanged.

For the purpose of sexual reproduction, haploid germ cells must be formed, namely sperm in the male and oocytes in the female. In principle, a diploid cell can be converted into two haploid cells by division without prior DNA replication. Each human germ cell would thus receive one chromosome 1 through 22 and one sex chromosome: X if a female diploid cell divided and either X or Y if a male cell divided. In reality, diploid cells about to give rise to germ cells replicate their DNA but then divide two times in succession to produce four germ cells. This specialized form of cell division is called **meiosis** (see Fig. 2.4) and occurs exclusively in the sex organs to produce germ cells.

During both mitosis and meiosis chromosomes become greatly condensed. This reduction in size is simply a packaging arrangement that compresses large DNA molecules into conveniently sized physical struc-

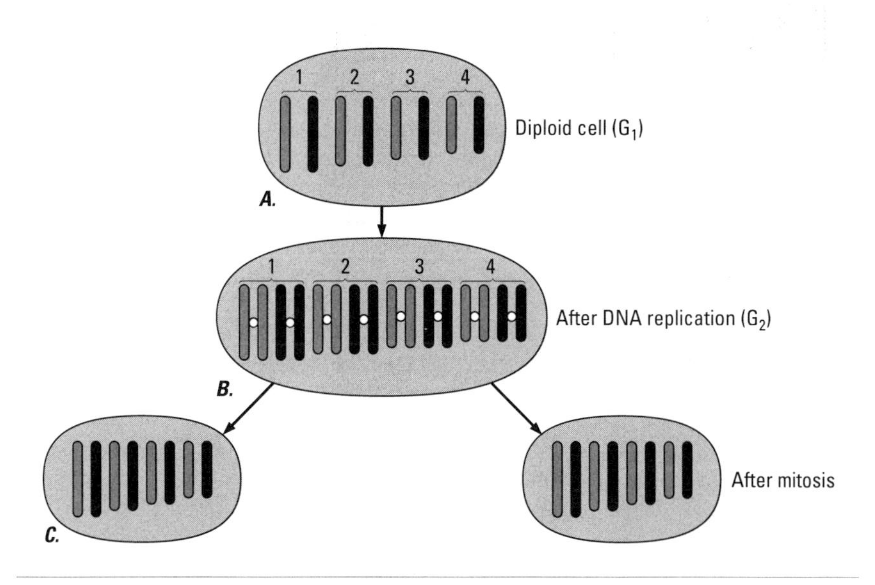

FIGURE 2.3 Mitosis. A hypothetical diploid cell with four chromosomes is shown. (*A*) G1 phase of the cell cycle, prior to DNA replication. For clarity, the two haploid chromosome sets inherited at the previous sexual generation from each of the parents are distinguished by color. (*B*) G2 phase, following DNA replication. The replicated chromosomes remain attached by a special structure called the **centromere.** (*C*) Daughter cells after mitosis. During mitosis, the centromeres separate and the replicated chromosomes are equally partitioned. In this way, the maternal and paternal complements of the genome remain unchanged from one division to another.

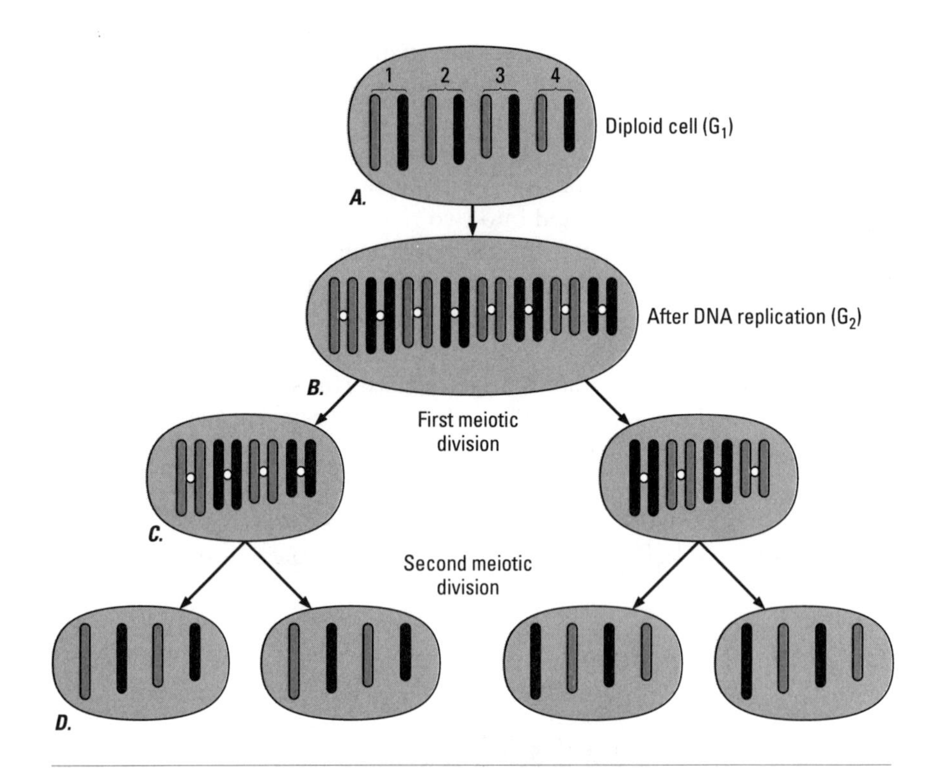

FIGURE 2.4 Meiosis: Hypothetical diploid cell with four chromosomes. As in Figure 2.3, the maternal and paternal chromosome sets are distinguished by color. Preparation for meiosis is identical to that for mitosis. Thus, *A* and *B* are identical to Figure 2.3. (*A*) G1 phase of the cell cycle, prior to DNA replication. (*B*) G2 phase, following DNA replication. The crucial difference between meiosis and mitosis is that meiosis involves two successive partitioning episodes of chromosomes. (*C*) First meiotic division. The distinction from mitosis is that the centromeres do not become separated. Instead, the replicated maternal and paternal chromosomes are partitioned into the two daughter cells. Herein lies the mechanistic basis for the principle of independent assortment, namely, that the partitioning of the chromosomes is random. Thus, the two daughter cells at the end of the first meiotic division are still diploid but are no longer carbon copies of the parental cell. For example, the left daughter contains two copies of chromosome 1 but both are maternal (green) in origin. Likewise, the right daughter cell contains only paternal versions of chromosome 1. (*D*) Second meiotic division. This division is identical to a mitotic division; the centromeres are separated and the chromosomes are partitioned equally. The daughter cells at the end of the second meiotic division are haploid and can go on to become germ cells, either oocytes or sperm.

tures; in this way, the mechanical task of partitioning the two genomes into two daughter cells is greatly facilitated. Mitosis and meiosis are but brief interludes in the life of a cell. During other portions of the cell cycle chromosomes are much less condensed.

The principle of independent reassortment states, simply, that during meiosis chromosomes are parceled out regardless of their origin (maternal or paternal) in the previous sexual generation. In other words, a particular germ cell may receive maternal chromosome 1, paternal chromosomes 2 through 4, maternal chromosomes 5 and 6, and so forth. A division of another diploid cell would produce a completely different assortment. The net result of independent reassortment is to break up the genome into 24 parcels (22 autosomes and 2 sex chromosomes) that can be shuffled at each generation. These parcels, however, are still very large. Each chromosome contains many genes, which, by virtue of residing on a single DNA molecule, are in effect physically linked and cannot be reassorted during sexual reproduction.

Homologous Recombination

The mechanism of recombination allows reassortment at the sub-chromosome level. On a conceptual level, the beauty and elegance of recombination is indeed striking. Consider a simple hypothetical cell, containing only one chromosome, which in turn contains only ten genes, designated A through J (see Fig. 2.5). The cell is diploid and contains a chromosome inherited from its father (black) with the genes ABcDEfGhIJ and a chromosome inherited from its mother (green) with the genes ABCDEfghIJ. Note the upper-case and lower-case symbols, which designate that the corresponding genes are not exactly identical: For example, gene c could be a mutant form of gene C by virtue of a single point mutation. The cell contains identical copies of genes A, B, D, E, f, h, I, and J; these genes are said to be **homozygous.** The cell contains, however, different copies of genes C and G: c and G on the paternal chromosome, C and g on the maternal chromosome. These genes are thus said to be **heterozygous.**

If the cell pictured in Figure 2.5 underwent meiosis, two of its germ cells would inherit genes c and G, while the other two would inherit genes C and g. What if, however, the combination C plus G was particularly advantageous for survival? On paper, the solution is simple: Both chromosomes can be broken at identical locations, parts exchanged, and the breaks resealed. As incredible as it may sound, that is essentially what

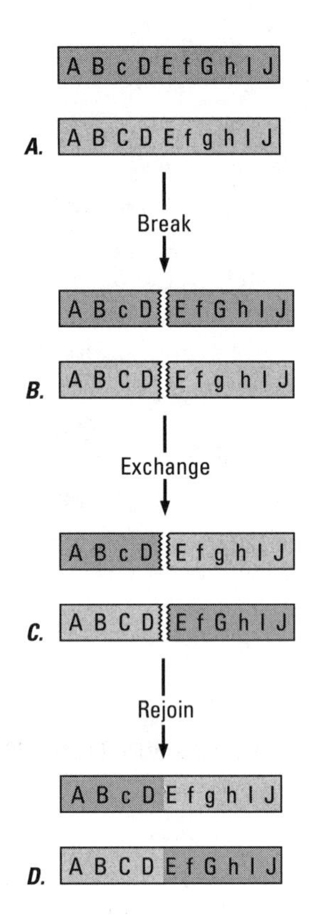

FIGURE 2.5 Genetic consequences of homologous recombination. The letters correspond to genes. The actual molecular mechanisms of homologous recombination comprise a number of discrete steps that lead to the final outcome (*D*).

happens in living cells. Although the precise molecular mechanisms are intricate and complex, the final outcome of **homologous recombination** is exactly as pictured in Figure 2.5.

Homologous recombination must be precise down to the nucleotide level; otherwise mutations would be introduced at the sites where chromosomes were broken and rejoined. The point where the chromosomes are broken, exchanged, and rejoined is often referred to as the **crossover point.** Genetic information is exchanged reciprocally between maternal and paternal chromosomes distal to a crossover point; it must be stressed, however, that no new information is created, either at the crossover point or anywhere else. Since there is no real physical limit to the number of crossovers that can occur in a particular stretch of DNA, the power and flexibility of homologous recombination as an agent of information exchange is indeed spectacular.

Two amendments must be introduced to this idealized picture. First, not all recombination events inside the cell are homologous. In certain cases DNA duplexes can be broken and rejoined at nonhomologous positions. Although some nonhomologous events have specific roles in the life cycles of certain organisms, most are random. In other words, such events are, in fact, mistakes that lead to mutagenesis through the rearrangement of DNA. In all probability they occur by mechanisms quite distinct from those involved in homologous recombination. Second, not all homologous recombination events are reciprocal. Instead of genetic information being exchanged between the homologous chromosomes, both chromosomes can be "converted" to contain either the maternal or the paternal sequence. In other words, genetic information is transferred in one direction only. This form of homologous recombination is called **gene conversion.** Although it may seem that reciprocal recombination and gene conversion are distinct, mechanistically speaking they are in fact just two manifestations of the same process.

Homologous recombination is catalyzed inside the cell by a host of specialized enzymes. If its only role was in sexual reproduction, one would expect a strict association with meiosis. However, homologous recombination activity is found, albeit at a lower level, in all mitotically dividing cells. Its function in this situation is almost certainly as a repair process. Since most mutations arise as errors during the replication of DNA, only one of the newly replicated DNA double helices will be mutant. Thus, the mutant DNA can be conveniently repaired by gene conversion with the second, unmutated duplex. This form of mutation repair is often referred to as postreplication repair and is strongly correlated with homologous recombination activity.

3

HOMOLOGOUS RECOMBINATION: MOLECULAR MECHANISMS

It is fun to sit down and imagine on paper how homologous recombination may occur. From such model building studies it quickly becomes apparent that the problem can be reduced to four distinct operations.

1. Search: Homologous sequences on separate chromosomes find each other and are brought into close physical proximity.
2. Breakage: DNA molecules are broken at homologous positions on the two chromosomes.
3. Exchange: DNA molecules are exchanged between homologous chromosomes.
4. Closure: Integrity and continuity of the two chromosomes are restored.

How can such operations actually occur at the molecular level? What kinds of enzymatic activities may be required? Do all the steps have to be, strictly speaking, catalyzed by enzymes?

The Molecular Paradox

The most mysterious of the four operations is the first one: How can homologous sequences find each other in a mammalian nucleus that contains 6 billion (6×10^9) base pairs of DNA? It is evident that homology must be defined operationally as base pairing between the two polynucleotide strands in the DNA duplex. There is really no other way that one piece of DNA can be compared to another and the question asked, Are they homologous?

If two homologous single-stranded DNA molecules are simply mixed in solution, they will in time find each other and base pair to form a double-stranded DNA duplex. If two nonhomologous single-stranded DNA molecules are similarly mixed, they will not pair. However, if the same two homologous molecules are mixed in their double-stranded forms, they will remain intact and separate forever. This is, of course, because the duplex form of DNA is quite stable. To separate the two strands requires a considerable amount of energy.

The crux of the dilemma is, How can duplex DNA molecules be compared for homology? Some reduction to single-stranded form, even if partial, is likely to be required. This can be brought about in several ways (see Fig. 3.1). For example, a portion of one polynucleotide chain could be removed from the middle of a DNA duplex to produce a single-stranded gap. A portion could be removed from the end of a DNA duplex to produce a single-stranded tail. A portion could be displaced and the gap filled by DNA synthesis to produce another kind of tail. Such single-stranded

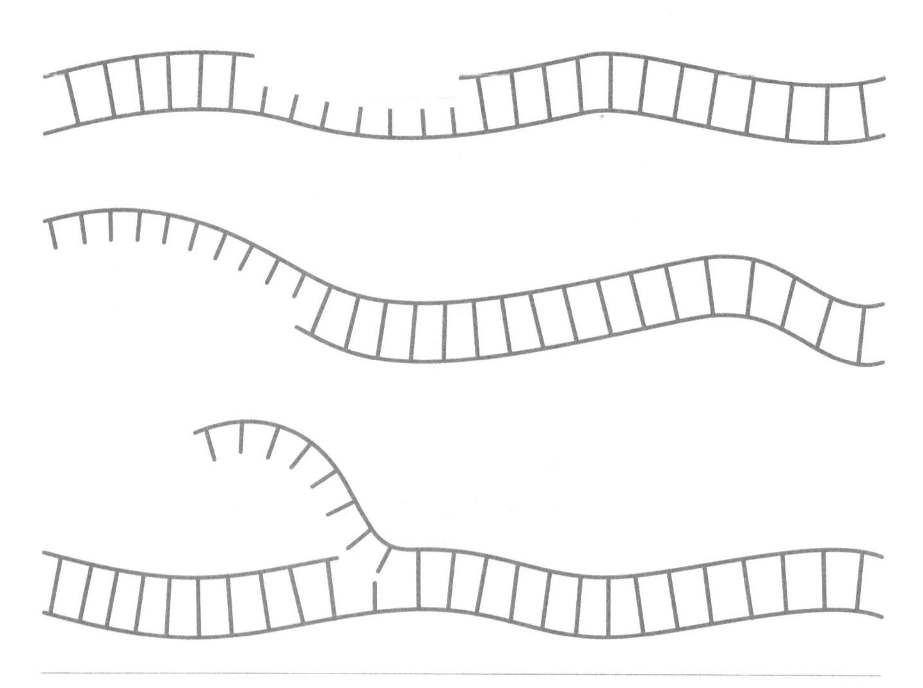

FIGURE 3.1 Substrates for the initiation of homologous recombination.

regions in DNA duplexes can be viewed as probes that are ready to test other DNA molecules for homology. Biochemical examination of living cells has indeed revealed the existence of numerous enzymatic activities eminently suited for the task of producing such single-stranded regions.

The paradox of the homology search is not resolved by introducing the concept of single-stranded recombinogenic substrates. From a mechanistic standpoint, to initiate recombination, such single-stranded regions would have to exist at identical positions on both homologous DNA duplexes in order to base pair. Indeed, early models of recombination postulated exactly that. The problem, conceptually, is twofold: Without benefit of a priori base pairing interactions, how does the cellular machinery identify homologous sites on two different duplex DNA molecules, and how do the single-stranded regions actually find each other?

The Molecular Structure of Chromosomes

Already in the 1920s microscopic studies revealed that homologous chromosomes become closely aligned, side by side, during meiosis. More recently, this morphological dissection has been greatly extended with the aid of the electron microscope. This alignment, however, cannot act as a guide in precisely defining homologous sites on separate DNA molecules. The reason is the architecture of the chromosome.

An average mammalian chromosome as present in meiosis or mitosis is about 3 to 5 μm long and contains approximately 150 million (1.5×10^8) base pairs of DNA. Considering the structure of DNA (Chapter 1), it becomes apparent that a molecule of that size has an overall end-to-end distance of approximately 5 cm. Obviously, the DNA must be very intricately packed to fit into an object 10,000 times shorter. This is achieved in three steps. First, the DNA duplex is wound around the outside of disk-shaped structures called **nucleosomes** (see Fig. 3.2*A*). Nucleosomes are made of protein molecules called **histones.** The DNA duplex makes nearly two full turns around each nucleosome, resulting in a fivefold compaction factor. Second, the nucleosomes, with the DNA duplex wound around them, are stacked in a helical array (see Fig. 3.2*B*). This results in a further eightfold compaction factor (an overall factor of fortyfold). The resultant protein–DNA structure, called **chromatin,** is a fiber of about 300 Å in diameter. At all stages of the cell cycle except meiosis and mitosis, DNA is found in the form of chromatin. Chromatin is a dynamic structure, becoming partially dissassembled and unwound

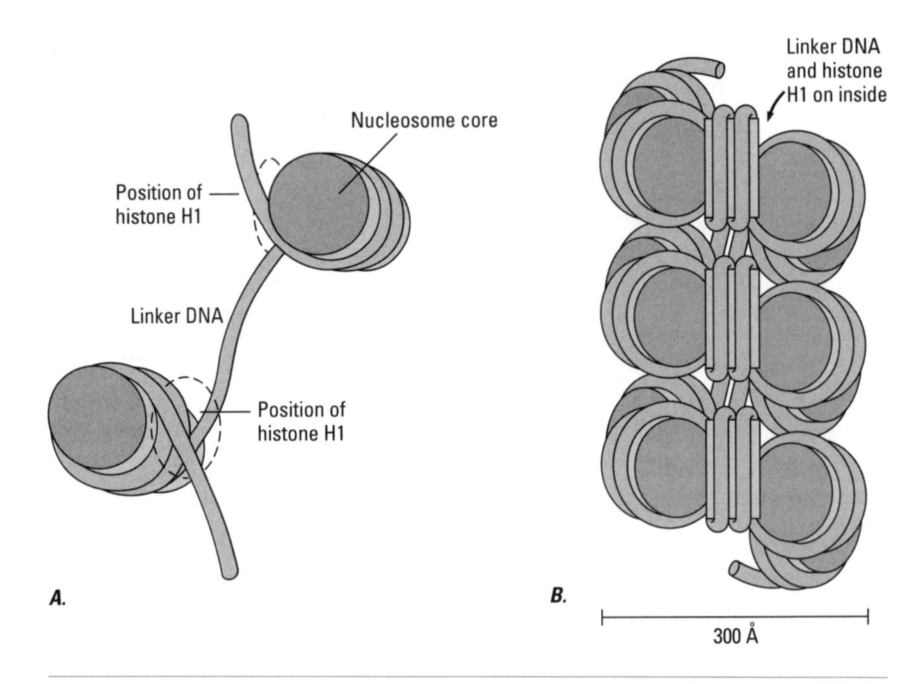

FIGURE 3.2 Structure of nucleosomes and their packing in chromatin. Chromatin contains five distinct types of histones, designated H1, H2A, H2B, H3, and H4. The disk-shaped nucleosome core particle around which the DNA is wrapped is composed of eight histone molecules, two each of H2A, H2B, H3, and H4. (*A*) The extended configuration, sometimes referred to as beads-on-a-string chromatin, represents the first stage of compaction. (*B*) The compact configuration, sometimes referred to as the 300-Å chromatin filament, represents the final stage of compaction. Histone H1 is not part of the nucleosome core, but it is essential for the formation of the 300-Å chromatin filament. It is positioned at the site where the DNA enters and leaves the nucleosome. In addition, H1 is a very elongated molecule, containing two arms that extend in either direction along the linker DNA. In the 300-Å filament, the molecules of H1 are found on the inside, where they contact each other closely and together with the linker DNA form a central core structure.

during transcription of genes into mRNA as well as during replication of DNA. Such local unwinding allows enzymes access to the DNA.

Obviously, further compaction has to occur in building a fully condensed chromosome. This is achieved by arranging chromatin in large loops, which are attached to a protein core called the **metaphase scaffold** that forms the center of the chromosome (Fig. 3.3*A*). Just prior to the first meiotic division, the scaffolds of the replicated maternal and paternal chromosomes become aligned to form a structure called the **synaptonemal complex** (Fig. 3.3*B*). It appears from microscopic studies that

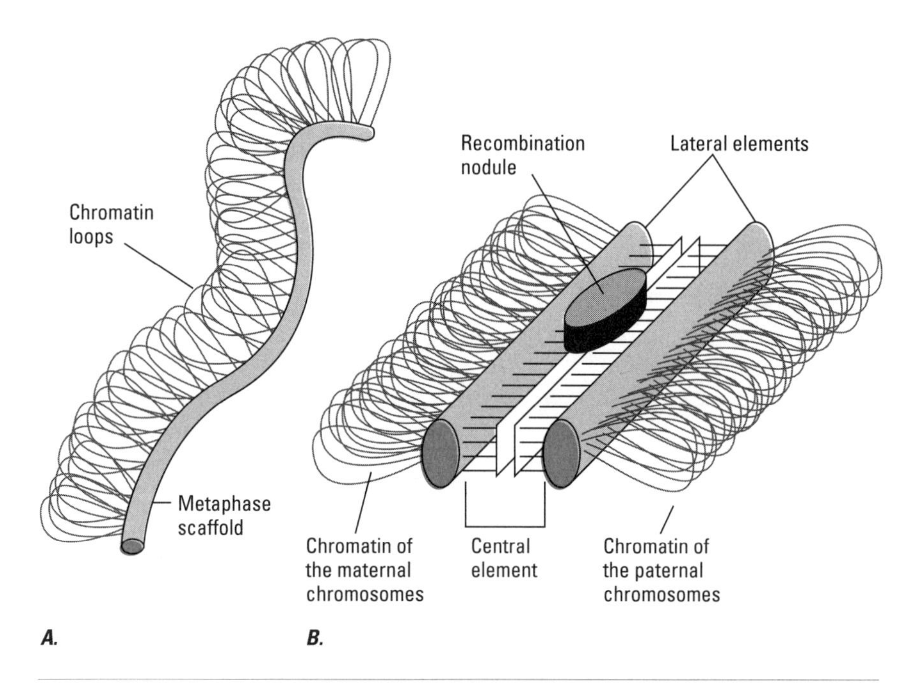

FIGURE 3.3 Packing of chromatin in chromosomes. (*A*) Portion of a condensed chromosome. For clarity, the chromatin loops are pictured pushed to one side to reveal the metaphase scaffold. The scaffold normally forms the center of the chromosome, with the loops radiating from it in all directions. (*B*) The synaptonemal complex is a special structure that forms during meiosis just before the first division. It is precisely at this time that homologous recombination between the maternal and paternal chromosomes is suspected to occur. Four chromosomes are involved in the formation of one synaptonemal complex, namely, the replicated maternal and paternal homologs (see Fig. 2.4*B*). The metaphase scaffolds of the two maternal and two paternal chromosomes fuse to form the lateral elements. The lateral elements are held together by a specialized structure, designated the central element, which has a ladderlike appearance with a distinct core element. This picture has been assembled almost entirely from electron microscopic observations. The lateral and central elements are known to be predominantly proteinaceous, but their exact composition, unlike that of the nucleosome, has not yet been elucidated. A globular structure, called the recombination nodule, is also often observed. Its function in the recombination process is not known; likewise, its specific molecular composition is a mystery.

the alignment is mediated by protein–protein interactions. In fact, the chromatin loops are seen to be separated on either side of the protein ladder forming the core of the synaptonemal complex. At this level of analysis however, occasional DNA–DNA interactions based on base pairing would not be detected. Nevertheless, it appears clear that the alignment of chromosomes does not result in a precise alignment, nucleotide by nucleotide, of the constituent DNA double helices. One possible idea would be to have the formation of recombinogenic substrates initiated at specific, predetermined positions on both chromosomes. Such sites of recombination initiation would have to be spaced relatively frequently and at more or less equal intervals, since a large body of data from genetic crosses in a wide variety of species indicates that crossovers can occur at any position along a chromosome. The single-stranded regions would then only have to find themselves by diffusion through a somewhat limited space, constrained by the higher order alignment of the chromosomes.

Recombinase Enzymes

An important advance in the understanding of homologous recombination was the discovery in the late 1970s of a protein that promotes a **strand exchange** reaction between a single-stranded DNA molecule and a homologous DNA duplex (Fig. 3.4). The protein is the product of the *E. coli recA* gene. From genetic studies, the *recA* gene was known for a long time to be absolutely required for homologous recombination.

It must be cautioned that most of what is known concerning the molecular details of recombinase action comes from studies of the *E. coli* RecA protein. Although more general features of recombination are likely to be conserved among all organisms, as is the case, for example, with translation, transcription, and DNA replication, many important differences in specific details will undoubtedly be found. It is thus somewhat dangerous to extrapolate too readily from bacterial to mammalian systems. Nevertheless, the body of knowledge concerning the *E. coli* RecA protein forms a meaningful framework and, significantly, at the present time the only one upon which speculations and investigations of recombination mechanisms in other organisms must be based.

The depiction in Figure 3.4 shows a test tube reaction using relatively small DNA molecules that are 100 percent homologous. Under these conditions the strand exchange reaction can proceed to completion, such that one DNA strand of the starting duplex molecule is entirely displaced and the single-stranded molecule is incorporated in its place. The duplex DNA molecule formed as a product of the reaction is referred to as a **heteroduplex,** since each of its strands is derived from a different molecule. Detailed experiments in the last few years have shown that the overall reaction can be divided into two parts: an initial homology recognition reaction that occurs over rather limited regions, sometimes referred

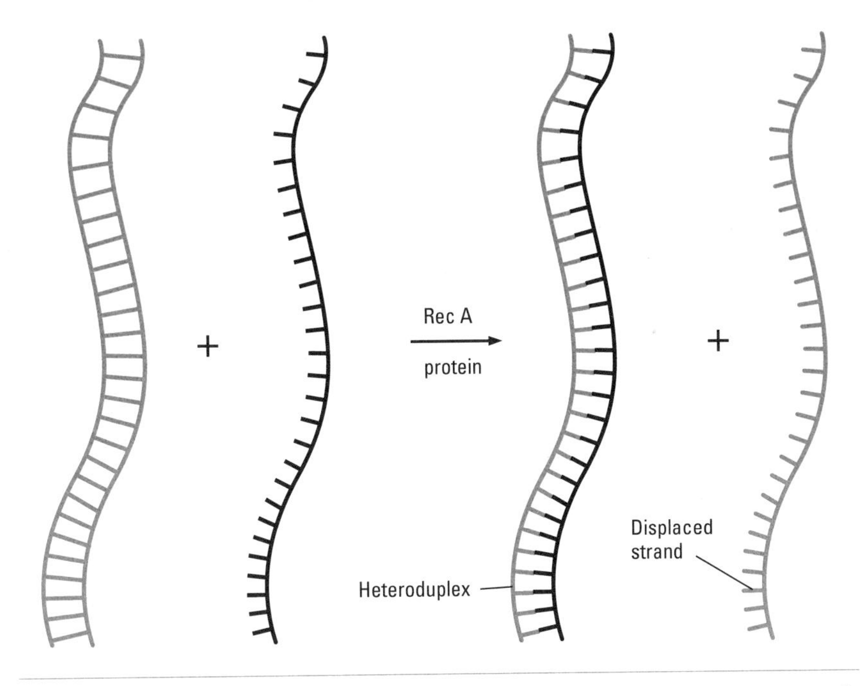

FIGURE 3.4 Strand exchange reaction. Only three components are necessary for the illustrated strand exchange reaction to proceed to completion: the DNA substrates, the RecA protein, and ATP (adenosine 5′–triphosphate). ATP is the main energy currency of the cell. When the terminal phosphate bond, often referred to as a **high-energy bond,** is broken to release ADP (adenosine 5′–diphosphate) and free phosphate, the chemical energy stored in the bond is released. The cleavage of ATP, often referred to as hydrolysis, can be coupled to other reactions in the cell. In this way the energy stored in ATP can be used to drive reactions that require a *net* input of energy. The RecA protein hydrolyzes ATP as it performs strand exchange, but it is not precisely known at what stage of the reaction ATP is used, or to what purpose it is put.

to as **strand invasion,** and the actual strand exchange reaction that follows, which can involve the entire molecules. In the living cell, the strand exchange reaction probably never proceeds to completion; heteroduplexes involving whole chromosomes have not been observed. Strand exchange can, however, occur over regions that are several thousand base pairs long, which is larger than the regions involved in the initial homology recognition.

The existence of recombinase activities greatly simplifies the homology search paradox: A single-stranded recombination substrate has to be found on only one of the two homologous DNA duplexes undergoing recombination. It is not difficult to imagine how strand invasion and exchange would operate in a more real-life setting, for example, on some of the substrates shown in Figure 3.1. The paradox, however, is not entirely solved. Rather, it is moved to a different level: How does the RecA protein actually perform the homology search and strand exchange reactions?

RecA Protein Coats Single-stranded DNA to Produce a Nucleoprotein Complex

It is not surprising that the RecA protein binds to DNA. The affinity for DNA is so high, in fact, that in a reaction containing unlimited amounts of RecA protein, all available DNA becomes completely coated. The nature of this interaction is quite curious. RecA protein and DNA become organized into a highly regular, helical structure (see Fig. 3.5*A*), sometimes called a **nucleoprotein filament.** A nucleoprotein is a generic designation for a complex between nucleic acid and protein. The nucleoprotein helix is right-handed, has a diameter of 100 Å, a pitch (rise per turn) of 95 Å, and contains six RecA monomers per one turn. One monomer of RecA protein is bound to three adjacent nucleotides of DNA. It is thus easy to calculate that one turn of the nucleoprotein helix consists of 18 nucleotides of DNA. These coordinates are only approximate but are probably correct within about 10 percent.

This picture of the RecA–DNA nucleoprotein complex has emerged primarily from high-resolution electron microscopic studies. Some intriguing and tantalizing shapes have been visualized (see Fig. 3.5*B*), but the level of resolution is insufficient to determine details of the interactions between RecA monomers or the exact location of the DNA in the complex. Model building studies suggest, however, that the DNA is probably located close to the inside of the helix. It must be emphasized that the affinity of RecA for DNA is not sequence specific; in other words, a nucleoprotein filament can be formed on any DNA molecule.

Although many of the mechanistic details of how the RecA protein acts to initiate homologous recombination remain to be elucidated, careful in vitro studies over the last ten years allow a generalized picture to be presented. It is intriguing that under certain in vitro conditions RecA protein will form complexes with both single-stranded and double-stranded DNA, and the coordinates of the resultant nucleoprotein helices are almost identical. The structure of duplex DNA must thus be substantially altered in the nucleoprotein complex; whereas the base-pair-to-base-pair distance in a free DNA duplex is 3.4 Å, this distance is increased to 5.1 Å in the nucleoprotein helix. This distortion of the DNA duplex may have a crucial role in the recombination mechanism. Under conditions that probably most closely resemble those found inside living cells, RecA protein binds preferentially to single-stranded DNA.

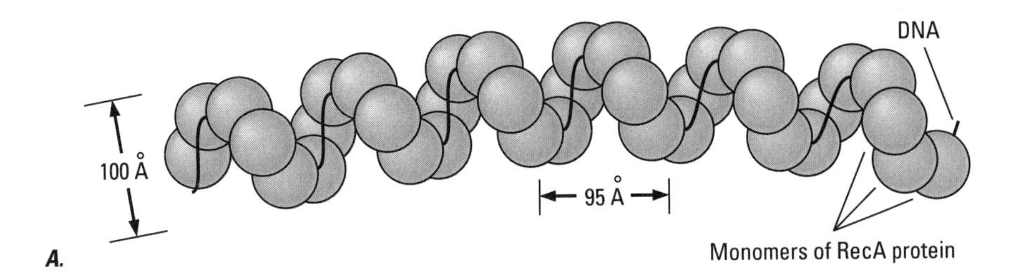

FIGURE 3.5 Structure of the RecA–DNA nucleoprotein complex. (*A*) Schematic representation of the RecA nucleoprotein complex. Each molecule of RecA is illustrated as a sphere. The DNA is shown as a thin black line running along the inside face of the RecA helix. (*B*) Computer reconstruction of the RecA nucleoprotein complex from observations by high-resolution electron microscopy. At this level of resolution, the DNA cannot be seen. It is apparent, however, that the RecA protein is not a simple sphere but has a rather complex shape. Most prominent are the two lobes on each molecule of RecA facing the outside of the nucleoprotein helix. (*Photo courtesy* Journal of Molecular Biology, *Academic Press, Ltd., London*)

The Homology Search Occurs Through RecA Nucleoprotein–Naked Duplex DNA Interactions

In living cells, single-stranded DNA regions generated by other enzymes are probably rapidly assembled into RecA nucleoprotein complexes. After such a nucleation event, assembly of the nucleoprotein helix can spread into adjoining regions of duplex DNA. More important, the binding of RecA to single-stranded DNA unmasks its affinity for duplex DNA. This can be dramatically demonstrated in the following experiment. A small amount of single-stranded DNA is allowed to react with RecA until it is all completely assembled into nucleoprotein filaments; naked double-stranded DNA is then added under conditions such that it will remain free of RecA. The result is that multiple contacts are rapidly formed between the nucleoprotein filament and the free duplex DNA, producing a complex, randomly aggregated meshlike network. In other words, a nucleoprotein helix containing single-stranded DNA becomes sticky to duplex DNA.

Extensive aggregates are formed even if the single-stranded DNA in the nucleoprotein complex and the free duplex DNA do not contain any regions of homology. Under such conditions, however, the reaction is terminated at the stage of aggregate formation. If the two DNAs contain regions of homology, however, the reaction rapidly proceeds to the strand exchange stage and the formation of heteroduplex DNA. It seems evident that it is at the aggregate stage when the initial homology search occurs and that it is the nucleoprotein helix containing single-stranded DNA that is performing this search.

What are the mechanics of this search? At this level of analysis, unfortunately, current knowledge gets rather tenuous. It has been proposed that the RecA protein contains two **binding sites,** one for single-stranded and one for duplex DNA. The binding of single-stranded DNA can then be postulated to activate the duplex DNA binding site. Presumably, when duplex DNA binds to the second site, it is compared for homology to the single strand occupying the first site. How does this happen? The process, at least from model building studies, appears geometrically complex.

Interactions Between Two Helical Objects Are Geometrically Constrained

Before continuing with the discussion of the homology search, a little aside concerning **topology** is in order. Topology is defined as the "study of those properties of geometric figures that remain invariant under certain transformations" (*Webster's Ninth New Collegiate Dictionary*). The relevance of this topic to the issues at hand becomes obvious when one considers how two helices, namely, the RecA nucleoprotein helix and the DNA double helix, when brought into close proximity, can be aligned in register for a homology comparison. In the last few figures, DNA has been drawn for simplicity's sake as a ladder. If that were really the case, alignment could occur over infinite distances without any topological problems. The situation is, of course, more complicated.

Consider, for example, a circular DNA molecule. Drawn in the form of a ladder, the two strands can be separated (see Fig. 3.6*A*). If, however, a single turn of the DNA double helix is drawn into the picture, the two strands become intertwined (see Fig. 3.6*B*). In other words, the two circles are locked, like links in a chain. For each additional turn of the double helix the **linking number** increases by one. Considering that the DNA

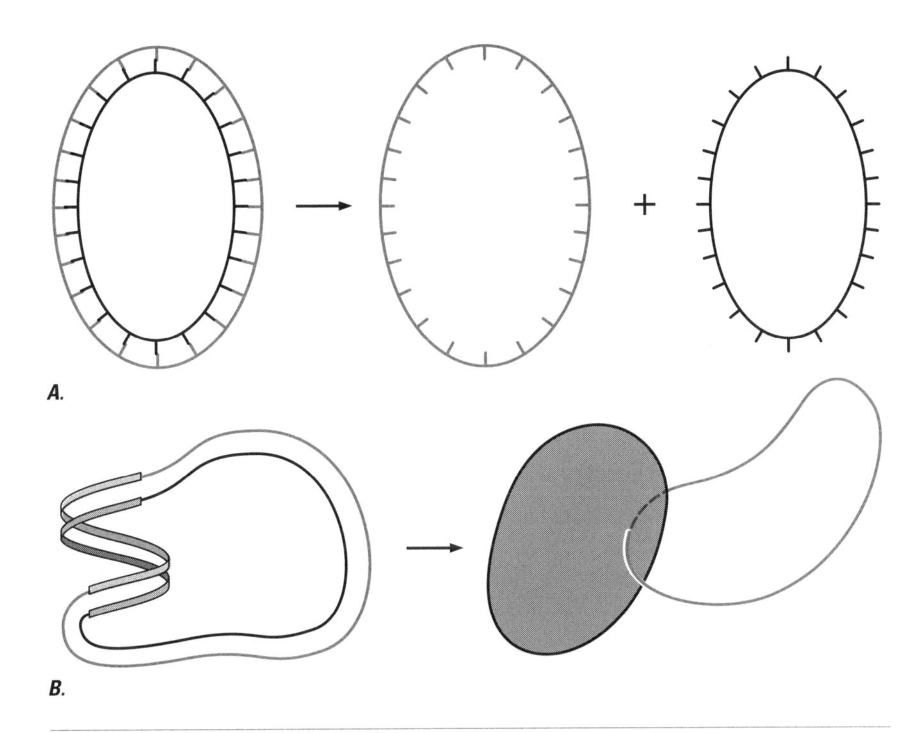

FIGURE 3.6 Linking of DNA strands in a double helix. (*A*) Circular double-stranded DNA molecule in the ladder representation. The linking number is said to be 0, because the two single strands can be separated by simply pulling apart without breaking either of them. (*B*) Same molecule with a single turn of the double helix (see Fig. 1.4) introduced into one portion of it. In addition, for simplicity's sake, the "rungs" of the DNA ladder have been left out. Any attempt to separate the strands by simply pulling them apart will reveal that the two circles are linked. Since one of the strands passes *once* through the *surface* defined by the other strand, the linking number assigned is 1.

duplex makes one turn every ten base pairs, it can be seen that linking numbers even in relatively small DNA molecules are very large.

As long as the two single strands remain continuous, the linking number is fixed and cannot change. If turns are taken out of one part, they will accumulate in another. This fact is well known to sailors and magicians. For example, if one pulls the two strands of the double helix apart in one portion of a circular molecule—in other words, decreases the linking number in that domain—the regions of DNA surrounding the bubble will rotate, and the linking number of the rest of the molecule will increase by a corresponding amount (see Fig. 3.7). It can be argued that in a linear molecule topological constraints will not be encountered, since the duplex regions surrounding an expanding bubble are free to rotate around their longitudinal axes. Unfortunately, chromosomes are long and intricately packed into small volumes. Free rotation of DNA is therefore physically impossible.

How can DNA molecules ever become unwound? One way would be to introduce swivels into duplex regions of DNA that surround expanding bubbles. Each turn of the swivel would allow the bubble to increase by ten base pairs (one turn of the double helix) without transferring the turns to the rest of the molecule. In other words, each turn of the swivel would decrease the linking number of the whole molecule by one. Such swivels, in fact, exist in all living cells. As may be expected, they turn out to be specialized enzymes. Such enzymes are called **topoisomerases,** and they catalyze the change in linking number of DNA molecules.

Topoisomerases do not actually allow free, unlimited rotation along the DNA axis. Being enzymes, they perform discrete operations in each catalytic step. Type I topoisomerases catalyze an *increase* in linking number in steps of *one.* The protein performs this operation by binding to duplex DNA, breaking one strand but remaining attached to both free ends, passing the other strand through the gap thus created, and finally resealing the gap (see Fig. 3.8). Type II topoisomerases catalyze a *decrease* in linking number in steps of *two.* These enzymes are somewhat more complex, but the essence of the catalytic mechanism is to introduce a double-stranded break in a duplex molecule, pass a neighboring loop of DNA duplex through the gap, and reseal the break.

The DNA double helix is rather stiff to rotation along its longitudinal axis. In other words, either overwinding or underwinding of the helix causes strain, and any further changes in the same direction tend to be resisted by the molecule. Topoisomerase I can act only in a passive mode to relieve strain in an underwound molecule by introducing positive changes in linking number. Topoisomerase II, on the other hand, can actively underwind a molecule by introducing negative changes in linking number. It is thus not surprising that the topoisomerase II reaction is

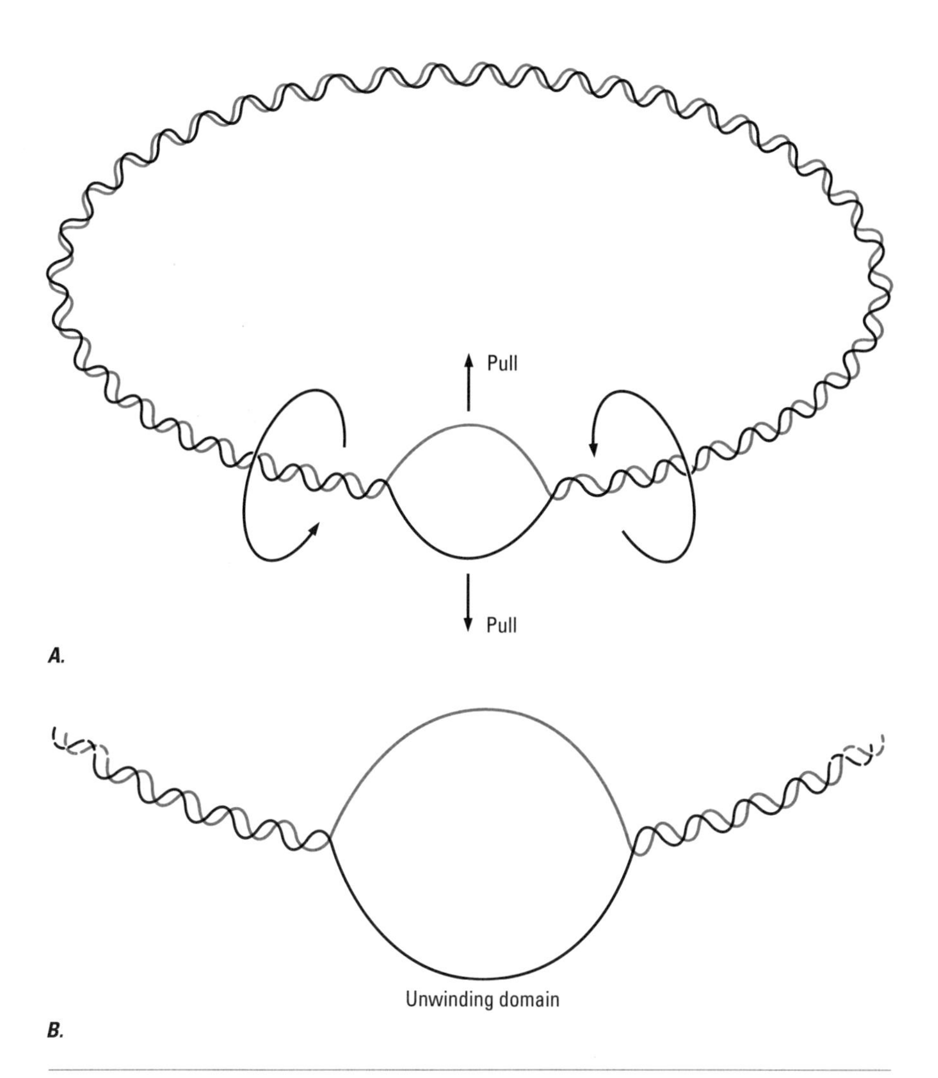

FIGURE 3.7 Unwinding of the DNA double helix in a topologically constrained molecule. (*A*) Circular double-stranded DNA molecule containing a single-stranded bubble, referred to as the unwinding domain. The size of the unwinding domain can be increased by pulling apart the single strands, as shown. This causes the duplex regions surrounding the bubble to rotate. The result is twofold: first, the bubble becomes larger and second, the rest of the molecule becomes more tightly wound. (*B*) Same molecule after one turn of the double helix has been unwound on each side of the bubble. The size of the unwinding domain has increased by 20 base pairs (two turns of the helix) and its linking number has remained at 0; the size of the rest of the molecule has decreased by 20 base pairs, and its linking number has increased by 2. The net result is that the linking number of the entire molecule has remained unchanged, since turns have simply been transferred from one region to another.

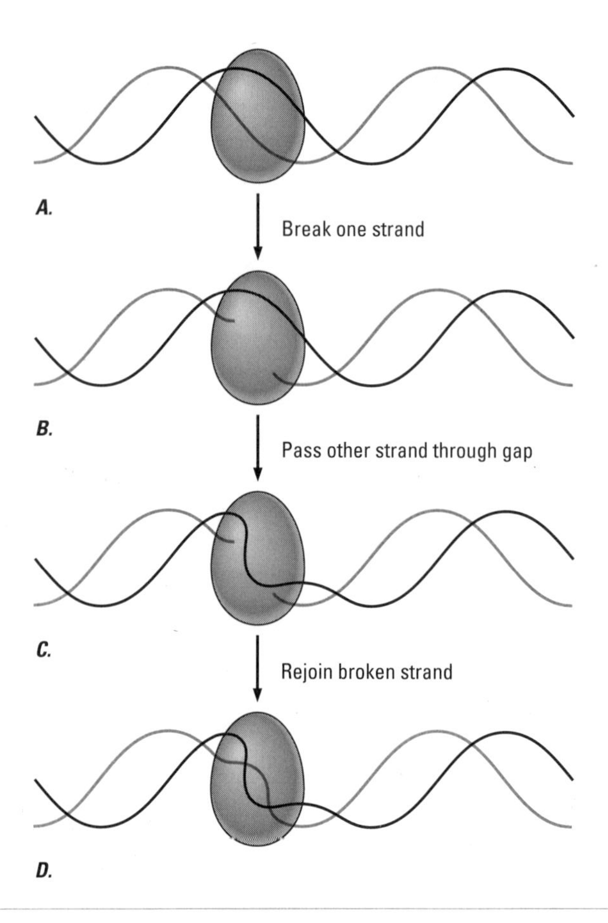

FIGURE 3.8 Catalytic mechanism of type I topoisomerases. (*A*) The enzyme is bound to a short portion of the double helix. (*B*) The enzyme has introduced a break in one of the strands. The enzyme becomes covalently attached to the 5′ end of the broken strand. In this way the energy of the broken DNA bond is preserved and can be used later to reseal the break. The enzyme also remains tightly attached to the 3′ end; in this way, free rotation of the DNA is prevented. (*C*) The enzyme passes the opposite, unbroken strand through the gap. (*D*) The enzyme rejoins the broken strand and finally dissociates from the DNA. The *net* result of the whole reaction is the increase of the linking number by 1.

coupled to hydrolysis of ATP, whereas topoisomerase I is ATP independent. The actions of the two types of topoisomerases are counterbalancing, and their relative activities need to be finely tuned inside cells. The bottom line is that between the two of them they can resolve just about any topological constraints that may arise in DNA, for example during recombination or replication.

Let's imagine how a close alignment between a duplex DNA molecule and its single-stranded homolog could occur in three dimensions. For a long time, it was believed that a likely mechanism would involve the unwinding of a portion of the duplex DNA molecule, perhaps with the aid of topoisomerases, to create a bubble. Next, a single-stranded tail or gap (see Fig. 3.1) would attempt to base pair with one exposed arm of the loop, again potentially utilizing topoisomerase activity to aid in the formation of an intertwined, helical conformation (see Fig. 3.9).

This picture is rather old-fashioned. First, it is now known that the single-stranded DNA would be coated with RecA protein to form a nucleoprotein filament. More important, recent experimental evidence indicates that single-stranded loops are probably not formed prior to the interaction of the RecA nucleoprotein complex with duplex DNA. Rather, double-stranded

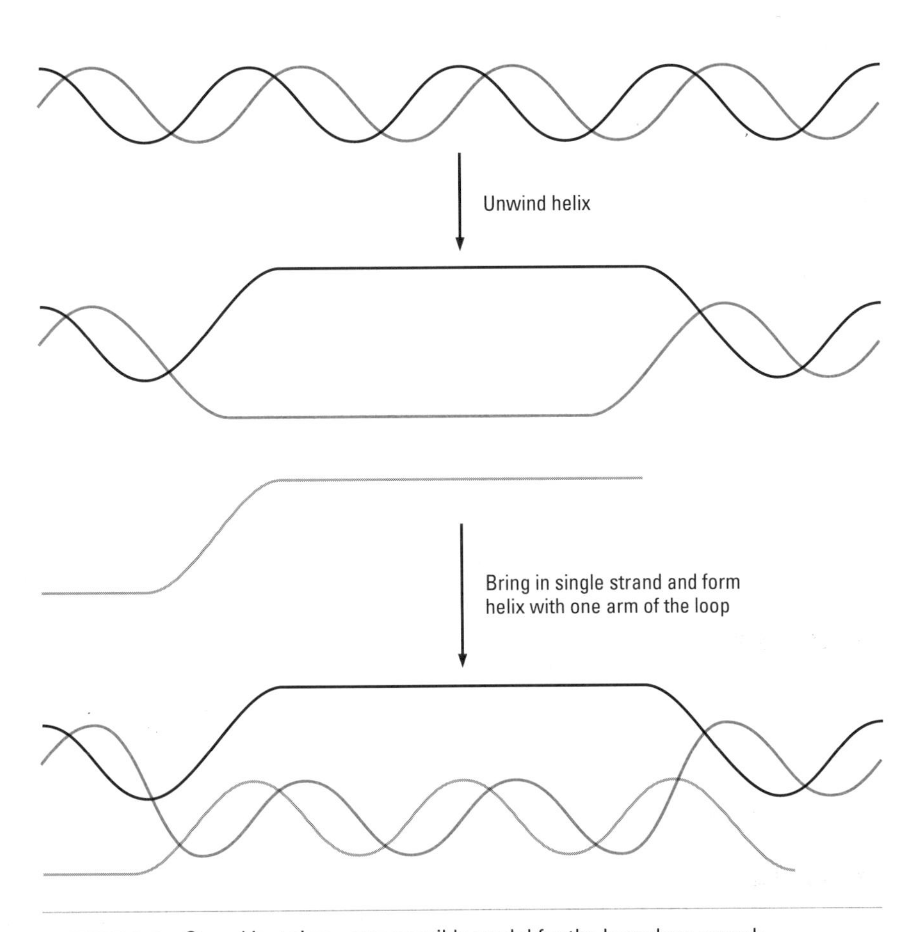

FIGURE 3.9 Strand invasion—one possible model for the homology search.

DNA seems to interact directly with the RecA nucleoprotein filament. The resultant interaction thus envisions two approaching helices that finally become intimately aligned and perhaps intertwined.

The Initial Homology Search Could Occur Through Very Short Stretches of DNA Sequence

Is this reasonable, and how much homology is likely to be enough? The key, of course, to the discussion at hand is the question, How many times can a particular unique, random target sequence be expected to occur in the genome just by chance? If homologous recombination could occur through sequences short enough likely to be present in multiple locations, the result would be the scrambling of the genome.

The occurrence of a particular sequence in a random stretch of DNA can be approximated using the equation $L = 4^n$, where n is the length of the specific target sequence sought and L is the minimum length of random DNA in which the target sequence has a significant probability of occurring at least once. For example, any particular unique six base-pair target sequence would be expected to occur approximately every 4000 base pairs ($4^6 = 4096$). Since the haploid mammalian genome is approximately 3×10^9 base pairs, the length of a target sequence expected to occur at least once can be calculated to be 15 ($4^{15} = 1.1 \times 10^9$). In fact, a 15-mer sequence would be expected, on average, three times in a haploid genome of 3×10^9; a 16-mer would not be expected to occur ($4^{16} = 4.3 \times 10^9$).

The problem with this calculation is that it is a gross oversimplification. Obviously, the genome is not a random string of nucleotides. Although a thorough discussion is beyond the scope of this book, the major issues should be mentioned.

Base Composition

In many organisms, the GC content of the genome can deviate significantly from 50 percent; this means that GC-rich target sequences are more likely to be found in GC-rich genomes and vice versa. The local base composition of genomes is also far from random; highly GC-rich as well as AT-rich regions are found interspersed throughout.

Gene Families

One often repeated theme in evolution is diversification of function; genes can be duplicated and then gradually evolved to assume new roles. Many genes can thus be grouped into families on the basis of sequence homology; a presumptive common ancestor is thus a logical postulate. The result is sequence similarity at various locations throughout the genome, depending, of course, on the amount of divergence since the original duplication.

Repetitive Elements

Mammalian genomes contain vast stretches, by some estimates up to 10 percent, of repetitive DNA. How did it get there? What is its function, if any? Repetitive DNAs can be grouped into several categories based on the length of the repeat sequence element, total copy number in the genome, and location. For example, some are composed of highly reiterated short sequences in one chromosomal location, others are longer and dispersed throughout the genome. The copy numbers can vary from a few to the hundreds of thousands. The occurrence of repetitive DNA is quite mysterious; a whole book could be written on this topic alone.

Initial Recognition Probably Demands Absolute Homology

The problem with repetitive DNA is, of course, that it complicates predictions about the minimal length of recognition sufficient for the initiation of homologous recombination. It can be calculated, for example, that a target sequence of 20 base pairs has a significant probability of occurring once in approximately 10^{12} base pairs ($4^{20} = 1.1 \times 10^{12}$), and that of 30 base pairs, once in approximately 10^{18} base pairs ($4^{30} = 1.2 \times 10^{18}$). These numbers appear high indeed compared to the size of the genome, but of course it is impossible to determine experimentally whether all possible 30-mers occur more than once in the genome. Undoubtedly, nature has settled the issue by trial and error over the eons of evolution, and the solution may well be a compromise.

Judging from genetic experiments in a variety of organisms designed to determine the minimum extent of sequence homology required for bona fide homologous recombination, it appears that the minimum recognition length may be somewhere between 30 and 40 base pairs. From the discussion above, it seems that 30 to 40 base pairs would give a safe enough margin as a homology tester. In vitro experiments with the RecA protein, although not completely unambiguous, also suggest that 30 to 40 base pairs are sufficient for a meaningful initial interaction.

Interestingly, the requirement for absolute sequence identity over this 30-to-40-base-pair stretch appears to be high. In other words, if two 30-mers were being compared for homology by the cellular recombination machinery and a single mismatch was found, the *net* result would be rejection for futher recombination. This makes sense with respect to the numerous repetitive elements; although they are highly repeated, their sequence conservation is not absolute and thus mismatches are likely to be detected and rejected by a strict homology criterion in the recombination machinery.

As is usually the case, no system is foolproof. In several instances, erroneous homologous recombination through repetitive sequence elements has been documented. Probably the choice in evolution has been to streamline and simplify the mechanisms of recombination without incurring too high a penalty due to an intolerable frequency of errors. On the other hand, some genome scrambling may even be beneficial from an evolutionary perspective.

The Alignment of the Two Helices Required for an Initial Homology Comparison May Be Achieved in Several Ways

The realization that an interaction spanning a contiguous stretch of 30 to 40 nucleotides may be sufficient for homology recognition is good news indeed! Since the RecA nucleoprotein helix contains 18 nucleotides per turn, such a distance would involve only about two turns of the helix.

For purposes of model building the duplex DNA can be considered as a flexible rod that can be wound around the RecA nucleoprotein helix. This is not an unjustified assumption, since any unwinding of the double helix as it is incorporated into the RecA nucleoprotein filament can be resolved by topoisomerases. The picture is complicated somewhat by the variety of single-stranded recombinogenic precursors (see Fig. 3.1). The range of possible interactions of each of these with duplex DNA is, from a topological standpoint, somewhat different. For example, if the interaction involved a single-stranded tail, either terminal or internal to the chromosome, the RecA nucleoprotein helix could work its way into the DNA duplex like a corkscrew. If the interaction involved a single-stranded gap internal to the chromosome, the duplex DNA could instead swing around the nucleoprotein helix as it was being taken up into it. This would not be possible, however, if the interaction took place at internal positions on both partners. In this case, a more complicated interaction involving topoisomerase action would have to be envisioned.

The above schemes require topological linking of the interacting helices. Intimate intertwining over short distances can be envisioned, however, without actually invoking any changes in linking number. Consider, for example, a loop of string being wound around a stick; in this analogy, the stick is a RecA nucleoprotein helix, and the string is naked duplex DNA (see Fig. 3.10). One arm of the loop of string is being wound in right-handed loops, whereas the opposite arm is being wound in left-handed

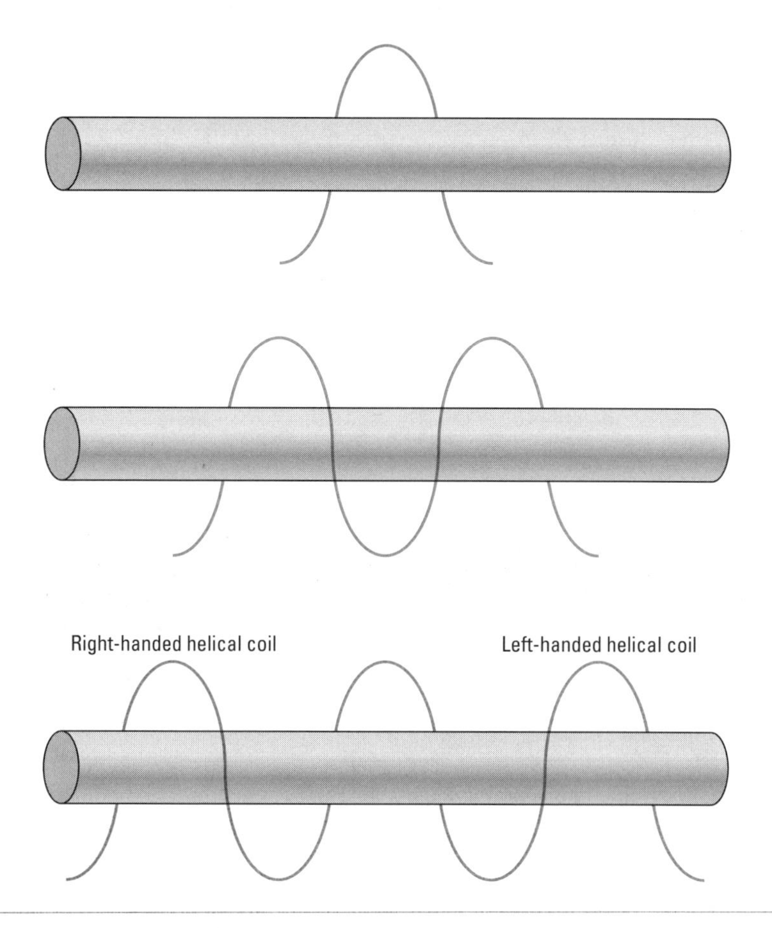

FIGURE 3.10 Helical interactions over short distances can occur without topological linking.

loops. The number of loops of each handedness is exactly equal, and thus the overall linking number between the two molecules, namely the stick and the string, remains zero. The picture can be easily redrawn in a more realistic mode for a possible interaction during initiation of homologous recombination (see Fig. 3.11). The arm of the loop that has the same handedness as the RecA nucleoprotein helix can be closely aligned with it, possibly leading to interactions culminating in actual homology comparisons. The other arm of the loop cannot interact, since its turns are opposite to those of the RecA nucleoprotein helix; it could be pictured simply wound very loosely around the circumference.

It is easily seen that too many turns would quickly result in a ball of knotted string, but two to three turns seem quite reasonable. Since two to three turns around the RecA nucleoprotein helix translates to 36 to 54 nucleotides worth of homology comparison, the overall picture looks feasible. The attractiveness of this model, from a theoretical aspect, is twofold. First, no topological linking of the two interacting partners needs to take place; thus, a single RecA nucleoprotein filament could perform many independent homology testing reactions in rapid succession. Second, an identical reaction mechanism could be used by all the different forms of single-stranded recombinogenic precursors (see Fig. 3.1).

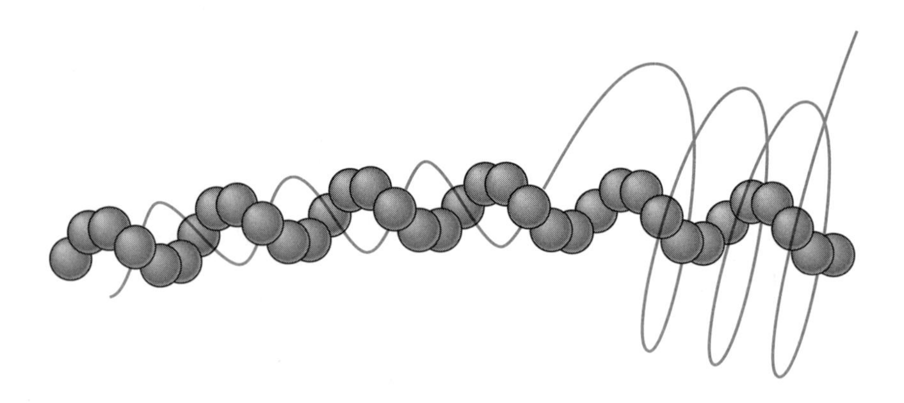

FIGURE 3.11 RecA nucleoprotein interacting with a naked DNA duplex without topological linking. A double-stranded DNA molecule, shown as a green string, is looped around a segment of a RecA nucleoprotein helix. At left, the duplex DNA is wound in right-handed coils and can interact closely with the nucleoprotein helix. At right, the duplex DNA is wound in loose left-handed coils. Three complete turns around the nucleoprotein filament are shown, but the two interacting objects are not topologically linked.

What happens, in what is likely to be the most frequent case, if homology is not detected? Is the duplex DNA released and the search continued randomly in three dimensions? Or does the duplex remain bound, and the search continue in one dimension by tracking, or scanning, along the bound duplex? Our knowledge concerning the detailed mechanics of the homology search is sketchy. This discussion demonstrates the complexities and suggests some possible theoretical solutions. To settle the issues experimentally has been difficult; from a technological perspective, some of the tools needed to ask the right kinds of questions are not yet available. In addition, it is almost certain that the formation of aggregates does not occur in the living cell in the same way it does in the test tube.

Homology Comparison Mediated by the RecA Protein May Occur by Formation of an Unusual Triple-stranded DNA Structure

How does the RecA protein detect homology between the two recombining partners? There is experimental evidence to suggest, as discussed above, that the RecA protein contains two binding sites, one each for single-stranded and duplex DNA. For a long time it was proposed that the RecA protein separated, or unwound, the duplex DNA into its two component strands. The single-stranded DNA in the other site would then be free to pair with one of the unwound strands in the duplex site. If pairing occurred and thus homology was detected, the recombination reaction could proceed to the strand exchange stage and beyond.

Unfortunately, an unwinding activity of the RecA protein has been difficult to demonstrate experimentally. Is there any way in which the homology comparison could be made without disrupting the base pairing in the duplex DNA? It has been known for a long time that under special conditions DNA can adopt configurations different from the standard B form (see Figs. 1.3 and 1.4), and some of these unusual forms can contain more than two polynucleotide strands. This is possible because the variety of hydrogen bonding interactions that can form between the bases is significantly greater than those found in the B form of DNA. For example, the N7 positions of guanine and adenine can be involved in additional hydrogen bonding interactions.

A triple-stranded form of DNA can form spontaneously in solution but only if the individual strands contain only purines or pyrimidines. Obviously, this limitation rules out a possible involvement of such structures in homologous recombination, where the interactions between polynucleotides of any sequence must be accommodated. Recently, however, it has been experimentally demonstrated that a triple-stranded DNA structure without any sequence restrictions can be formed by the RecA protein. Although this triple-stranded form of DNA will not form without the action of the RecA protein, once formed it is relatively stable and will persist even if the RecA protein is removed from the complex. It was this property that actually allowed the detection and preliminary characterization of this novel form of DNA. Significantly, its formation can be catalyzed by recombinase proteins purified from higher organisms, including mammals, as well as the *E. coli* RecA protein.

The precise structure and hydrogen bonding interactions of this novel triple-stranded form of DNA have not yet been experimentally elucidated. A possible structure based on model building studies has been proposed (Fig. 3.12) and contains the following features. The first and second strands form a duplex containing normal hydrogen bonding interactions characteristic of B form DNA and, in addition, bond with the third strand. The third strand can contain both purines and pyrimidines and is oriented parallel to its identical strand in the duplex. The third strand is located in the major groove of the duplex and forms hydrogen bonding interactions only with the purines in the duplex. In other words, the third strand alternates its pairing with the two duplex strands, depending on which strand contains a purine at a particular position.

Once homology is detected, perhaps via a triple-helix intermediate, can the strand exchange reaction proceed without unwinding of the DNA duplex? Again, the answer is not known, but a possible mechanism has been proposed on the basis of model building studies. The model envisions a RecA protein with two binding sites (see Fig. 3.13). Binding of single-stranded DNA into site I activates site II for binding of double-stranded DNA. As the double-stranded DNA enters into site II, its structure is distorted to match the nucleotide-to-nucleotide spacing of the single strand in site I, namely, the pitch of the double helix increases from 34 Å to 51 Å. It is possible that this extreme distortion of the DNA duplex is a key feature that allows the three-stranded structure to form. If homology is detected over a sufficient length, the RecA protein is activated into the strand exchange mode of action. In this phase of the reaction, the RecA protein catalyzes a **rotation** of the bases of the single-stranded DNA in site I and its complementary strand in site II downward into a correct orientation for the formation of hydrogen bonds leading to a B-form DNA heteroduplex.

It should be emphasized that although this model appears elegant and reasonable, it is still speculative. One curious fact about the activity of the RecA protein is that although the initial homology recognition is sensitive

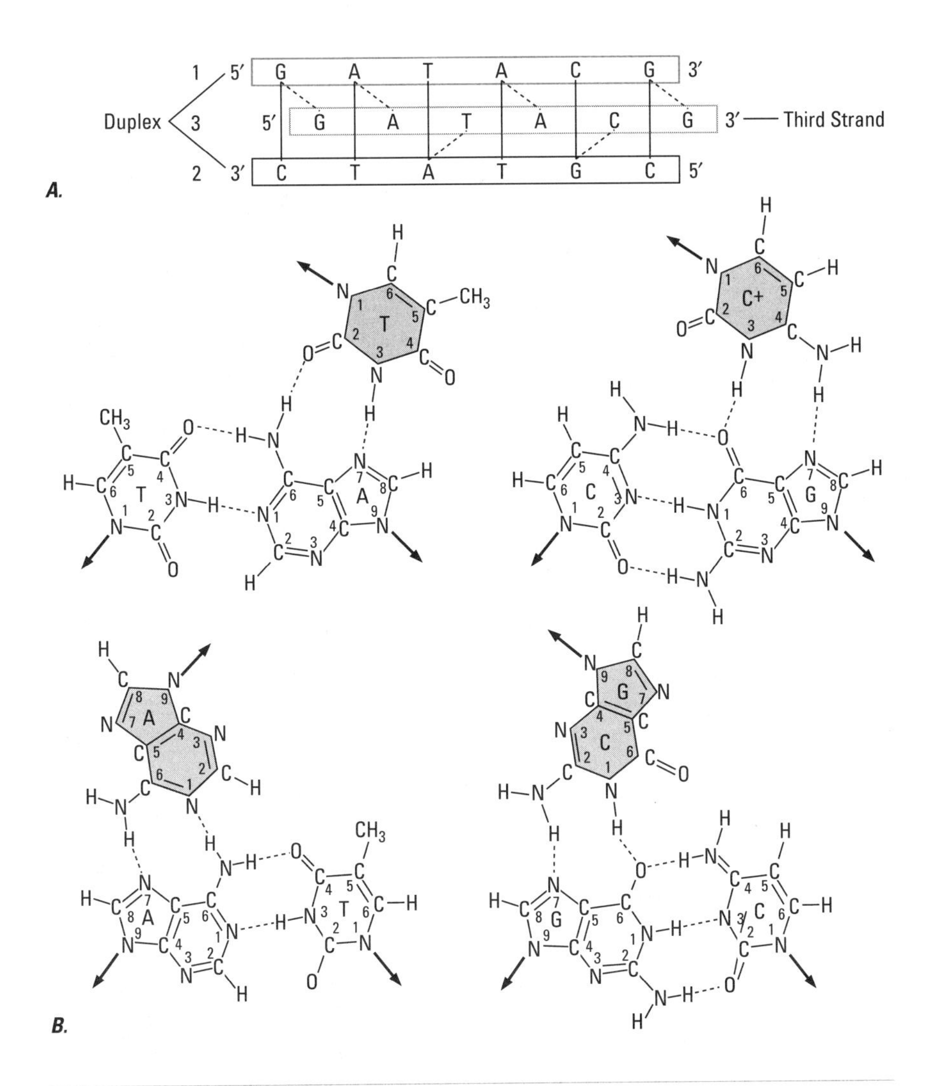

FIGURE 3.12 Putative pairing scheme for a DNA triple helix. (*A*) Proposed three-stranded structure. Strand 1 (dark green) and strand 2 (black) are held together by hydrogen bonding interactions found in the usual B form of DNA, indicated by solid black lines. Strand 3 (grey) has an identical sequence and is oriented with the same polarity as strand 1. Hydrogen bonding interactions between strand 3 and the other two strands are indicated by dashed lines. (*B*) Proposed hydrogen bonding interactions. Only the bases are drawn; the bonds to deoxyribose are indicated by an arrow. The bases in the third strand are shaded light green.

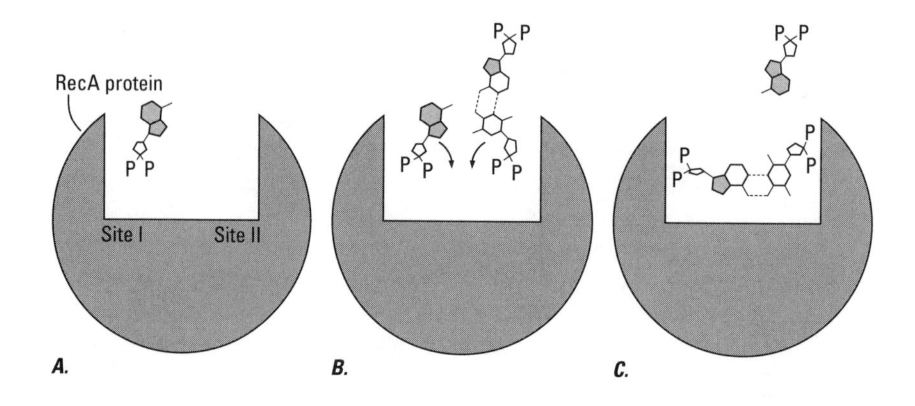

FIGURE 3.13 Strand exchange reaction envisioned as a rotation of bases in a two-site model of the RecA protein. The DNA is shown in transverse cross section, with one nucleotide of each chain and the 5′ and 3′ phosphate groups. The rings of the bases are shaded green; the deoxyribose ring is left open. Hydrogen bonds between the bases are shown as dashed lines. The drawing is to scale, except that the diameter of the RecA protein is 60 percent larger than shown. (*A*) Single-stranded DNA molecule entering site I. At this stage the protein would be activated for the binding of duplex DNA into site II. (*B*) Duplex DNA molecule having entered site II. At this stage the homology comparison would be executed via a three-stranded intermediate. If the outcome of the comparison was positive, the protein would be activated for the strand exchange reaction, which would proceed by the rotation of the bases downward as indicated by the arrows. (*C*) Newly formed heteroduplex occupying binding sites I and II. The displaced strand is leaving the nucleoprotein complex.

to mismatched base pairs, the actual strand exchange reaction can proceed past regions of substantial nonhomology. It is almost as if the RecA protein looked for a green light (the recognition of perfect homology over a sufficient distance), and once the switch was thrown, a new and different mode of operation took over (the strand exchange reaction). It is not known which step or steps along this complicated reaction mechanism is coupled to the hydrolysis of ATP.

Nucleases

The nucleases, a large class of enzymes, catalyze the hydrolysis of the phosphate backbone of nucleic acids. Nucleases can be subdivided into several categories according to their specificity of action. Some will attack both DNA and RNA; the ones of interest in recombination are specific for

DNA and are called deoxyribonucleases, or **DNases** for short. DNases can be further divided into two broad categories: **endonucleases** and **exonucleases** (see Table 1). Endonucleases can attack DNA anywhere along its entire length and introduce a break; exonucleases can attack only the ends of DNA and progressively chew their way into it. Some DNases are specific for either single-stranded or double-stranded DNA; others can attack both forms. Endonucleases that attack double-stranded DNA can be further subdivided into those that can break both chains of the DNA and those that can break only one chain (introduce a nick). Exonucleases can be subdivided by their specificity for the ends of DNA: Some will attack 3′ ends, while others prefer 5′ ends. Some nucleases can chew DNA all the way down to free mononucleotides, while others nibble it down to short pieces of a few nucleotides in length.

The variety of DNases is large, and they find functions in all aspects of DNA metabolism. Some are involved in the replication of DNA, others in DNA repair, or recombination. Still others are involved in the salvage and recycling of the building blocks of nucleic acids, or protecting the organism from assault by invading viruses or exogenous DNA. Many of the enzymes involved in various aspects of DNA metabolism can probably participate in more than one process. This may explain why, for example, *E. coli* mutants deficient in exonuclease VII have a hyper-recombination phenotype. Although the major function of the enzyme appears to be DNA repair, one consequence of its activity may be the degradation of single-stranded recombinogenic substrates generated by other enzymes. Absence of exonuclease VII may thus increase the intracellular concentration of certain recombinogenic substrates. Along the same lines, the activities of some of the nucleases overlap, and it seems reasonable that under some circumstances one enzyme may be able to substitute for another. This may explain why, for example, *E. coli* mutants deficient in exonuclease I do not show any overt phenotype.

Other Enzymes Involved in Homologous Recombination

Besides the recombinase activity required to pair homologous DNA sequences, other enzyme activities are required to generate recombinogenic substrates and to terminate and resolve crossovers. It is easy to imagine how some of the nucleases listed in Table 1, or even a combination of such nucleases acting in concert, could generate the various recombinogenic substrates illustrated in Figure 3.1. DNA polymerase

TABLE 1 NUCLEASE ACTIVITIES FOUND IN *E. coli*

ENZYME	DIRECTION OF HYDROLYSIS	SUBSTRATE	COMMENTS
Exonuclease I	3′→5′	Single-stranded	Degrades up to terminal dinucleotides
Exonuclease II (an activity of DNA polymerase I)	3′→5′	Single-stranded	Proofreading function during DNA synthesis
Exonuclease III	3′→5′	Duplex	Activity on duplex end as well as nick (also endonuclease II activity)
Exonuclease IV	3′→5′	Single-stranded	Complete degradation
Exonuclease V	3′→5′ 5′→3′	Duplex	Also endonuclease activity, no action at a nick, ATP-dependent, RecBC gene product, functions in recombination
Exonuclease VI (an activity of DNA polymerase I)	5′→3′	Duplex	May function in excision repair
Exonuclease VII	3′→5′ 5′→3′	Single-stranded	Produces oligonucleotides, mutants lacking enzyme have hyperrecombination phenotype
Endonuclease I		Single-stranded, Duplex	May function in salvage or protection from exogenous DNA
Endonuclease II		Duplex	Produces a nick next to apurinic site, repair function
Repair endonucleases		Duplex with lesions	Nick at 5′ end of lesion
Restriction endonucleases		Unmodified duplex	Protection from exogenous DNA, viruses

Source: Adapted from Arthur Kornberg, *DNA Replication* (New York: W. H. Freeman, 1980).

activity is required to fill in single-stranded gaps that may arise during recombination. Specific endonucleases are also involved in the resolution of crossovers. Finally, DNA ligase is required to seal any nicks that are left over and restore the integrity of the DNA duplexes.

One interesting enzyme that participates in homologous recombination is the single-strand binding, or **SSB,** protein. The function of the protein is to bind and prevent the reassociation of single-stranded regions of DNA. The SSB protein is known to function in several processes where single-stranded regions need to be stabilized, such as replication and repair of DNA, as well as in recombination. The SSB protein coats the DNA in a beads-on-a-string fashion, but although the binding is tight enough to prevent DNA base pairs from reforming, it does not prevent the interaction of DNA with other proteins such as the RecA protein or DNA polymerase.

One *E. coli* enzyme that generates single-stranded tails that serve as substrates in the initiation of homologous recombination has been studied in considerable detail. The enzyme is a complex of three distinct polypeptides encoded by the *recB, recC, and recD* genes. The RecBCD enzyme complex has several activities. First, like the RecA protein, it has ATPase activity. Second, it has an activity that can unwind duplex DNA into single strands; the hydrolysis of ATP is probably used to drive this reaction. Third, it has 3′→5′ and 5′→3′ exonuclease activity, as well as single-strand endonuclease activity. This multitude of activities made it somewhat difficult to figure out how the enzyme creates recombinogenic substrates inside the cell.

The reaction mechanism turned out to be quite remarkable (see Fig. 3.14). The RecBCD enzyme attaches to a double-stranded end and subsequently travels down the DNA, unwinding the duplex ahead of itself. As the enzyme moves unidirectionally along the DNA, it appears to interact with only one strand of the DNA. Since this strand is taken up by the enzyme faster than it is released, a continuously growing single-stranded loop is created and travels along with the enzyme. If the enzyme encounters and traverses, in the 3′→5′ direction, the nucleotide sequence 5′-GCTGGTGG, referred to as the **Chi site,** it cleaves the looped-out strand at its base and releases a single-stranded tail.

The RecBCD enzyme is not the only activity responsible for creating recombinogenic substrates in *E. coli;* for example, since it requires a double-stranded end to enter the DNA, it cannot act on circular DNA molecules. Other enzymes or combinations of enzymes are also active in the initiation of homologous recombination. The consideration of the RecBCD activity, however, establishes one very important concept: the existence of recombinogenic sequences. Genetic studies in a variety of organisms have for a long time implied the existence of recombinational **hot spots,** in other words, points along the chromosome where homologous recombination can be observed to occur at elevated frequencies with respect to neighboring regions. The study of the RecBCD enzyme has revealed one possible molecular mechanism to explain this phenomenon.

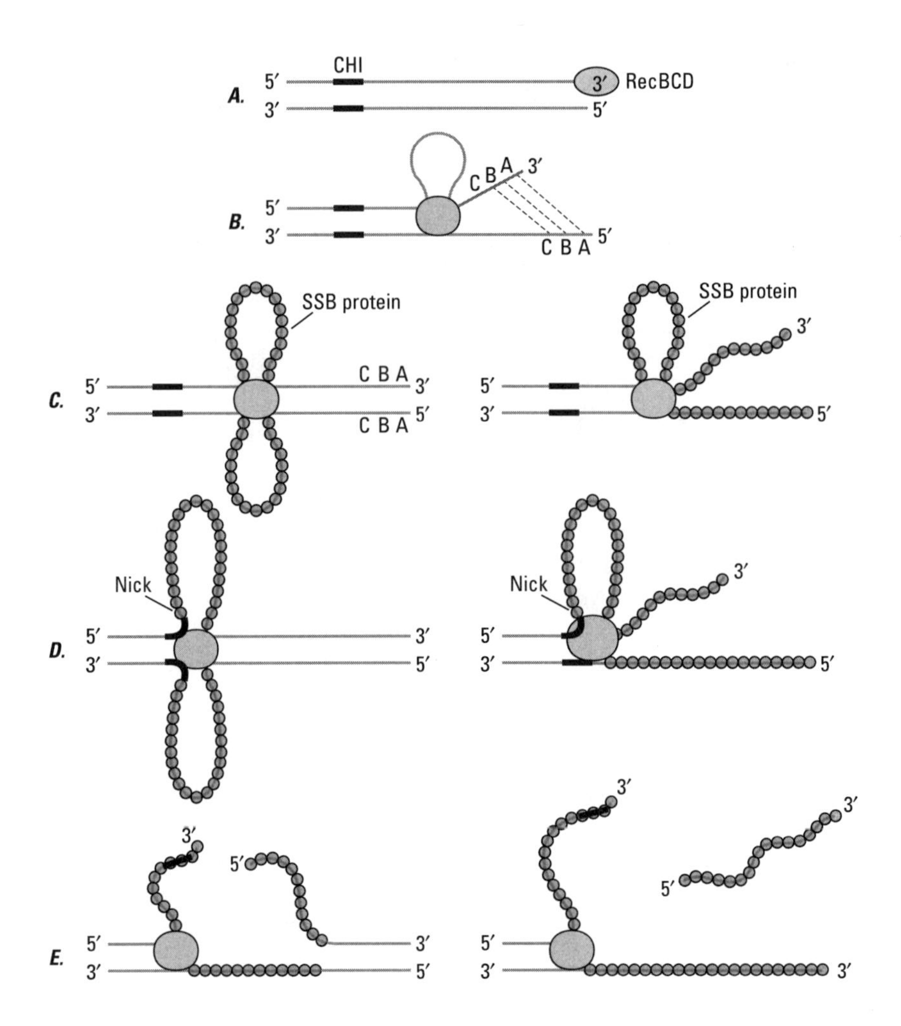

Suggested Readings

Howard-Flanders, P., West, S. C., and Stasiak, A. (1984). Role of RecA protein spiral filaments in genetic recombination. *Nature* 309: 215–220.

Cox, M. M., and Lehman, I. R. (1987). Enzymes of general recombination. *Annu. Rev. Biochem.* 56: 229–262.

Kucherlapati, R., and Smith, G. R., eds. (1988). *Genetic recombination.* Washington, D.C.: American Society for Microbiology.

FIGURE 3.14 (*Left*) Creation of single-stranded recombinogenic substrates by the RecBCD enzyme. (*A*) The RecBCD enzyme (light green) entering a free end of duplex DNA (grey). (*B*) A short region of DNA has been unwound. The top strand of the DNA is taken up by the enzyme at approximately 300 nucleotides per second but released at approximately 200 nucleotides per second. This causes the loop to grow at approximately 100 nucleotides per second. The emerging strand is thus complementary not to the opposite strand in the immediate vicinity of the enzyme but only to more distal regions (A, B, and C and dashed lines). (*C*) Two possible fates for the continuing interaction of the enzyme complex with the DNA. At left, base pairs have formed between the regions A, B, and C; the duplex has spread right up to the enzyme displacing a single-stranded loop on the bottom strand. The two loops are of equal size and are stabilized with SSB protein (dark green). As the enzyme complex moves forward, both loops grow at an equal rate, and duplex DNA is formed at the rear. This is referred to as the **twin-loop** mode of action. At right, SSB protein is shown binding to the single-stranded tails before they can base pair. As the enzyme complex moves forward, both the loop and the tails keep growing, but the continued binding of SSB protein keeps them all as distinct structures. This is referred to as the **loop-tail** mode of action. In vitro, high concentrations of SSB protein favor the loop-tail mode, and lower concentrations the twin-loop mode. It is not known which mode of action is predominant in vivo. (*D*) Enzyme complex encountering a Chi site (black). In both modes of action this activates an endonuclease activity that introduces a nick at the base of the top loop. (*E*) Enzyme complex has moved past the Chi site, releasing single-stranded tails that can subsequently be coated with RecA protein.

Smith, G. R. (1989). Homologous recombination in *E. coli:* multiple pathways for multiple reasons. *Cell* 58: 807–809.

Radding, C. M. (1991). Helical interactions in homologous pairing and strand exchange driven by RecA protein. *J. Biol. Chem.* 266: 5355–5358.

4

MODELS OF HOMOLOGOUS RECOMBINATION

Following the detailed look at the wheels and cogs of the recombination machinery in Chapter 3, it is time to back off and try modeling the whole process from beginning to end. Recombination has been studied for almost a century by geneticists using a variety of organisms as model systems. Any molecular model one wishes to put forth must thus attempt to satisfy this body of accumulated genetic data.

Explaining Gene Conversion Is the Real Testing Ground

A favorite model system for studying recombination is certain lower fungi, such as the molds *Ascobolus, Neurospora* and *Sordaria*, or the yeast *Saccharomyces*. One reason is that the life cycle of these organisms allows the recovery of all the products of a single meiosis. Diploid yeast cells can divide indefinitely by mitosis, but when subjected to adverse environmental conditions, such as limitation of certain nutrients, they undergo a distinct developmental program called sporulation (see Fig. 4.1). During sporulation, diploid cells enter meiosis, and the resultant four haploid cells encase themselves in thick protective coats and withdraw from the cell cycle. The four progeny haploid cells, now referred to as spores, remain bound together in a saclike structure called an ascus. Spores are very tough and can withstand adverse conditions such as desiccation for long periods of time. When suitable growth conditions again present themselves, the haploid spores shed their protective coats in a process called germination, reenter the cell cycle, and become normal vegetative cells. The resultant haploid cells can divide indefinitely by mitosis. To complete the life cycle, under certain conditions two haploid cells can fuse in a primitive form of mating to produce a diploid vegetative cell.

One difference in the life cycles of the molds is that the four haploid cells at the end of meiosis undergo a single mitotic division before being converted into spores; the ascus thus contains eight haploid spores. Each of the eight spores in the ascus is thus the descendant of one of the eight single DNA strands entering the first division of meiosis (see Fig. 4.2); since homologous recombination occurs at this stage, each of the partners in that process can be recovered as a pure haploid lineage (a clone) of organisms. In some molds, such as *Neurospora,* the spores remain ordered in a linear array in the ascus that reflects their genesis during meiosis.

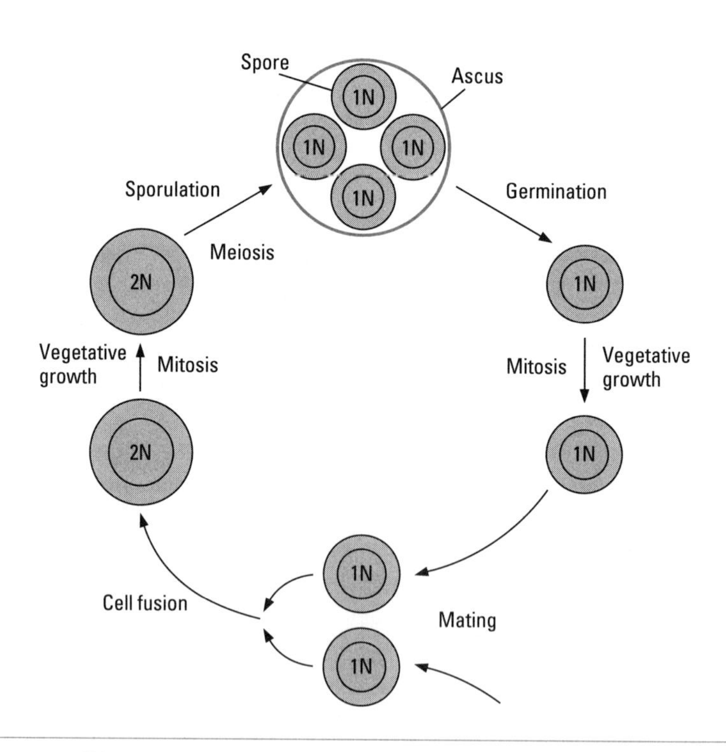

FIGURE 4.1 Life cycle of the yeast *Saccharomyces serevisiae*. The nuclei are shown in light green and the cytoplasm in grey. The ploidy of each nucleus is indicated. Cell walls and nuclear membranes are shown in black; the ascus wall is shown in dark green.

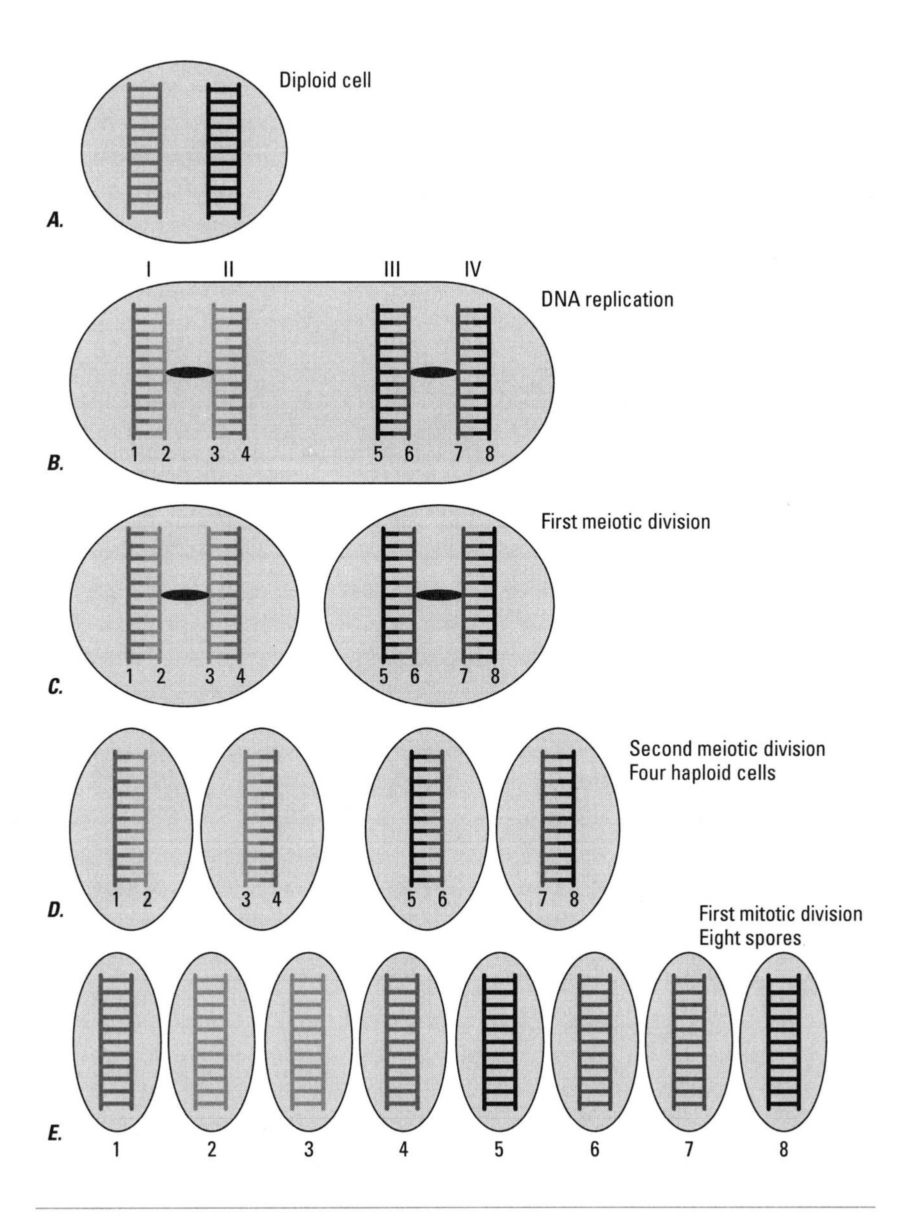

FIGURE 4.2 Segregation of DNA strands during meiosis in the mold Neurospora. (*A*) Diploid vegetative cell in G1 phase of cell cycle. For clarity, the haploid genome size has been reduced to one chromosome; the maternal and paternal homologs are shown in green and black. (*B*) Cell in meiosis just prior to first division. The newly synthesized DNA strands are indicated in light green and grey. The centromeres are shown as grey ovals. It is at this stage that homologous chromosomes are aligned to form the synaptonemal complex and recombination between maternal and paternal homologs takes place. (*C*) Products of first division. Note that the centromeres remain intact. (*D*) Four haploid cells resulting from second meiotic division. (*E*) Eight haploid spores formed from the four haploid cells by a single mitotic division. The spores are enclosed in a linear array in the ascus.

In a cross between two genes, for example AB × ab, recombination would yield the new combinations Ab and aB in the progeny. If the individual genes, A versus a or B versus b, were scored independently in the eight spored ascus, their segregation would be always 4 : 4, as long as the recombination event was reciprocal (see Fig. 4.3). This is, of course, because in a reciprocal cross genetic information is neither created nor destroyed; it is merely exchanged. At a low frequency, however, aberrant segregation, such as 6 : 2 or 5 : 3, is observed in genetic crosses. The appearance of such asci is the evidence for a unidirectional, nonreciprocal transfer of genetic information, namely gene conversion. At face value, it seems as if genetic information was being deleted from certain positions and replaced with information from the homologous position on another chromosome. The necessity to accommodate gene conversion in its various manifestations is the primary testing ground for all models of recombination.

The Holliday Model

The first generally accepted molecular model of recombination was proposed in 1964 by Robin Holliday (see Fig. 4.4). In brief, the model envisions the initiation event as a symmetric nicking of both partners at homologous sites, followed by mutual strand exchange on one side of the nicks resulting in the formation of symmetrical heteroduplex DNA. The crossed strands hold the two partners together in a structure called a **Holliday junction.** The Holliday junction can be resolved by nicking the appropriate single strands to result in either parental or recombinant configuration of flanking markers.

Although some aspects of the model, such as the symmetric initiation events, are almost certainly wrong, other features survive to this day and have been incorporated into more recent models. The most important advance was the notion of the Holliday junction (Fig. 4.5). Model building studies suggest that the crossed-strand connections can be made without disrupting the individual DNA duplexes. Base pairing and stacking can be maintained intact right up to the junction without chemical bond strain or unfavorable contacts. In fact, the crossover of a strand from one helix to another can be accomplished in a single repeat unit of the sugar–phosphate backbone.

As usually depicted, the Holliday junction contains two outside and two connecting strands. It is easy to see, however, that the four strands are in fact equivalent, since the positions of the strands can be exchanged by

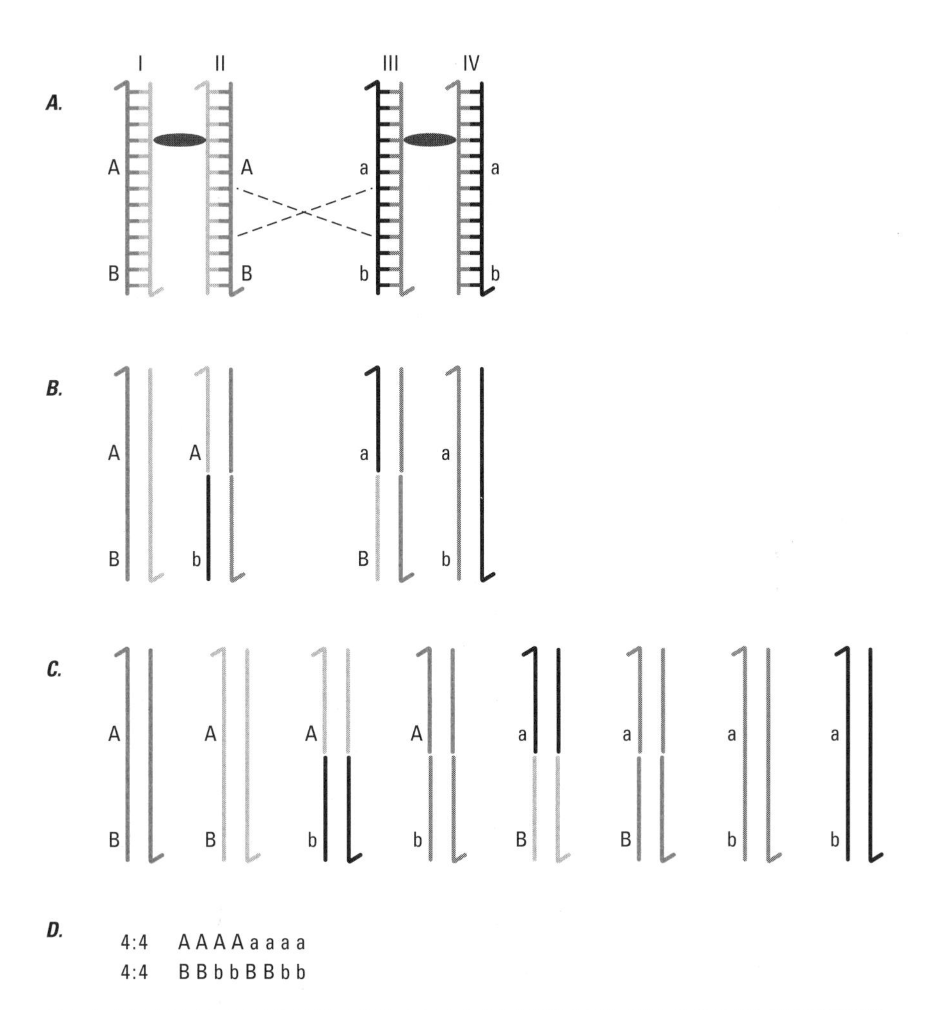

FIGURE 4.3 Segregation of genetic markers during reciprocal recombination in *Neurospora.* (*A*) Alignment of chromosomes in synaptonemal complex (as in Fig. 4.2*B*). The color scheme is the same as that in Figure 4.2. For clarity the chromosomes are numbered I to IV; I and II are maternal and III and IV are paternal in origin. The arrowheads indicate the locations of the 3′ ends of DNA strands. The dashed lines indicate the location of one possible crossover. Other crossovers, namely between chromosomes I and III, I and IV, or II and IV, are equally likely. Crossovers can also occasionally occur between two maternal or two paternal chromosomes, for example between chromosomes I and II, but in most circumstances the result of this recombination event would be genetically indistinguishable from no recombination having taken place, since both maternal chromosomes contain identical information. (*B*) Configuration of chromosomes immediately after the homologous recombination event. For simplicity, the ladder representation of DNA and the centromeres have been omitted. (*C*) Configuration of chromosomes in eight-spored ascus (as in Figure 4.2*E*). For intervening steps, refer to Figure 4.2*C, D.* (*D*) Segregation of genetic markers A/a and B/b in this genetic cross.

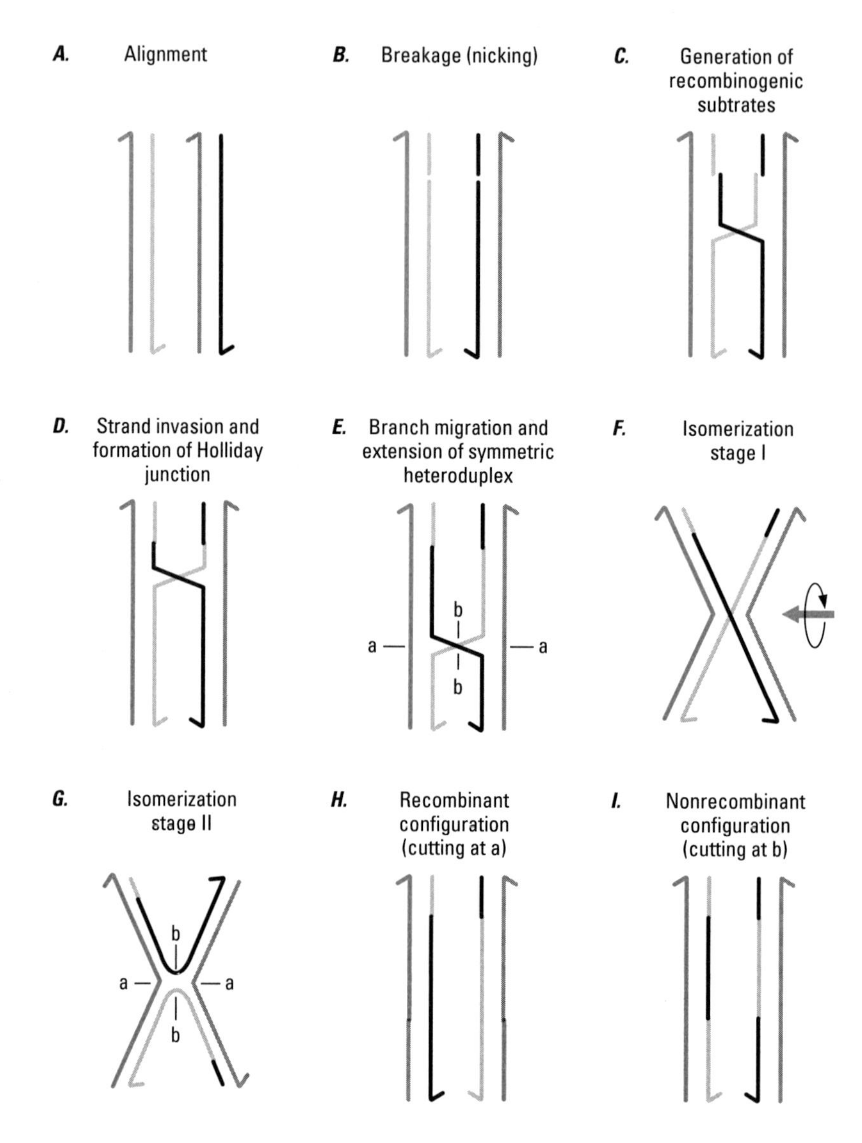

rotation of the duplex DNA arms of the junction around the central connection. The process of rotating the arms is often referred to as **isomerization** of the junction and is likely to be in a state of rapid equilibrium at physiological temperatures (see Fig. 4.4*F*, *G*). It is one of the key features of the model, since it predicts an equal probability of generating either parental or recombinant configurations when the junction is terminated by endonuclease cleavage.

FIGURE 4.4 (*Left*) Holliday model. Refer to Figures 4.2 and 4.3 for color scheme. Only the two chromosomes involved in recombination are shown (I and IV in Fig. 4.3). (*A*) Chromosomes aligned prior to recombination. (*B*) Nicks are introduced in strands of the same polarity at homologous positions on both chromosomes. The right chromosome has been rotated 180° along its longitudinal axis for clarity in illustrating further steps. (*C*) The duplex neighboring each nick is unwound to create a single-stranded tail and a gap. The tails are seen approaching the partner chromosome. (*D*) Each tail base pairs with the complementary gap on the opposite chromosome, creating a Holliday junction (see also Fig. 4.5) flanked by symmetrical heteroduplex regions. The heteroduplexes are said to be symmetrical because their location and extent are identical on both chromosomes. (*E*) The Holliday junction has branch migrated towards the bottom (see also Fig. 4.6), extending the heteroduplex regions created by strand invasion. Branch migration can proceed in either direction; movement towards the top would terminate the Holliday junction. (*F,G*) Isomerization of the Holliday junction. In stage I, the arms are drawn outwards to create a figure X configuration. In stage II, the rightward arms have been rotated 180° around a horizontal axis. (*H*) Resolution of Holliday junction resulting in reciprocal crossover event. This is achieved by introducing two nicks at positions marked a in E or G. (*I*) Resolution of Holliday junction not resulting in reciprocal crossover event. This is achieved by introducing two nicks at positions marked b in E or G. Although the chromosomes return to their parental configuration, regions of symmetric heteroduplex are left between the sites of initiation and termination of recombination.

The equivalence of the strands in the Holliday junction leads to the other key feature of the model, **branch migration.** In this process, the two identical bases above and below the crossover exchange places (see Fig. 4.6). The exchange is thermodynamically neutral, since for every base pair broken a base pair is formed. The crossover point can readily diffuse along the joined duplex DNAs in a rapid, zipperlike fashion. Branch migration can thus be viewed as a passive process leading to the formation of heteroduplex DNA. Large heterologies between the two recombining duplex DNAs, such as deletions or insertions, will block branch migration; point mutations, such as single mismatches and frameshifts, have little or no effect.

Holliday junctions have been well studied using in vitro model systems, and their existence in living cells has been demonstrated by several methods. Not surprisingly, Holliday junctions appear in all models of recombination proposed to date. As discussed in Chapter 3, heteroduplex DNA can also be formed by the direct action of strand exchange activities of recombinase proteins, and in this case strand exchange can proceed past regions of significant heterology. It is not clear to what extent the formation of heteroduplex DNA in living cells is due to passive branch migration or to active strand exchange promoted by recombinase proteins.

FIGURE 4.5 Space-filling model of two DNA duplexes joined by a Holliday junction. (*Photo courtesy* Journal of Molecular Biology, *Academic Press, Ltd., London*)

FIGURE 4.6 (*Right*) Branch migration of Holliday junctions. (*A*) Branch migration using the ladder representation of duplex DNA. In the top drawing, the crossover junction is located between nucleotide positions 5 and 6. The homologous nucleotides immediately to the right of the crossover (position 6 in each duplex) are highlighted. If the top nucleotide abandons its partner, swings down, and base pairs with its complement in the homologous duplex, and the bottom nucleotide performs the analogous operation in the opposite direction, the crossover junction will move to the right by one position, as shown in the bottom drawing. (*B*) Topologically more realistic three-dimensional representation of branch migration. All four arms of the junction are geometrically equivalent. One easy visualization of the branch migration process is to consider the crossover junction, referred to in the figure as the exchange site, to be fixed in space. The two interacting DNA duplexes are then rotated in one direction around their longitudinal axes. In the figure, a right-handed rotation is depicted (black arrows), causing the arms on the right side of the junction to progressively shorten and the arms on the left side of the junction to lengthen by a corresponding amount (green arrows).

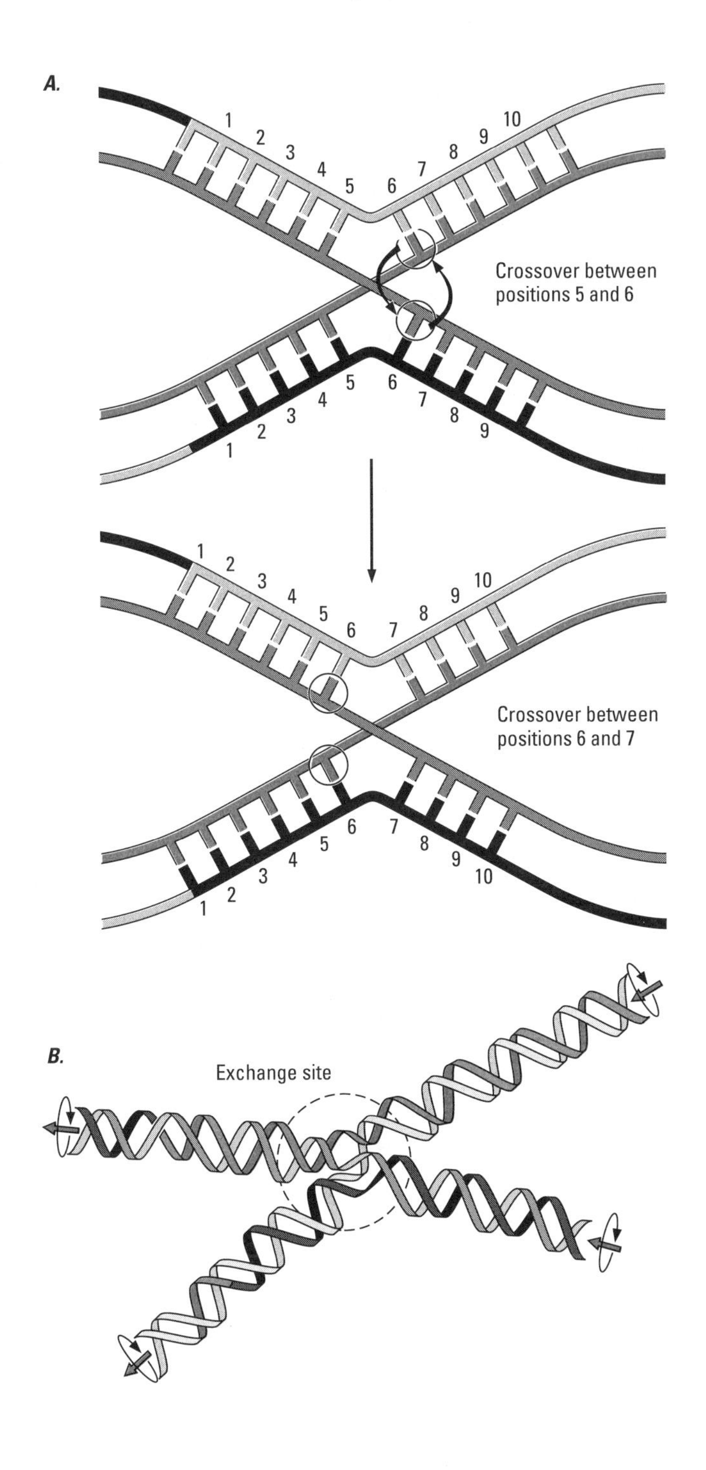
A.
1
2
3
4
5
6
7
8
9
10
Crossover between
positions 5 and 6
Crossover between
positions 6 and 7
B.
Exchange site

Repair of Heteroduplex DNA Can Explain Gene Conversion

Since formation of heteroduplex DNA can extend across regions of imperfect homology, the question arises, What happens to the resultant duplex that contains mismatched bases or other kinds of lesions? Mutations arise continuously by such processes such as oxidative damage to DNA and replication errors. Not surprisingly, elaborate enzymatic machineries have been evolved to repair such lesions. With respect to the fate of heteroduplex DNA generated during recombination, the obvious question is, Which strand is used as the template for repair? The rules of the repair game are elaborate, but under some conditions, particularly those of meiotic recombination, either strand can act as the donor of repair information. On the other hand, if a mismatch is not repaired prior to the next round of DNA replication, the information in each strand of the duplex is simply passed on to one of the daughter cells.

The generation of heteroduplex DNA, followed by repair or lack thereof, is the key postulate of the Holliday model that explains gene conversion. Consider, for example, a cross in which three neighboring genes, ABC × abc, are being scored (see Fig. 4.7). Recombination, initiated during

FIGURE 4.7 (*Right*) Repair of mismatches in heteroduplex DNA. The color scheme is the same as in Figures 4.2, 4.3, and 4.4. (*A*) Homologous chromosomes aligned prior to recombination. Three genes, A, B, and C, are shown on the maternal chromosome (green). The paternal chromosome carries a mismatch mutation in each of the genes, marked a, b, and c to distinguish them. (*B*) Crossover between genes A and B. (*C*) Holliday junction migrated past gene B. (*D*) The mismatches in the regions of heteroduplex DNA are indicated with bulges. (*E*) One possible outcome of mismatch repair: both of the chromosomes encode maternal (*B*) information. At the end of meiosis, each of the chromosomes is partitioned into one of the four haploid cells. This accounts for two of the four cells, since only the chromosomes involved in recombination are shown (I and IV of Figs. 4.3 and 4.4). The other two chromosomes continue to carry only maternal or paternal information. (*F*) Result of the one mitotic division that follows meiosis. Below the drawing is a summary of the genetic makeup of the spores resulting from this sequence of events. The spores containing recombinant chromosomes are boxed. (Since the segregation of chromosomes during cell division is random, other arrangements of spores are possible. Referring to the numbering of chromosomes in Figure 4.3, the linear order of chromosomes shown in this figure is II–I–IV–III. A chromosome order of, for example, I–II–III–IV is equally probable and would result in the pattern AAAAbbbb, BBBBbbBB, ccccCCCC. The overall genetic ratios, however, remain unchanged in all possible configurations, namely, 4A : 4a; 6B : 2b; 4C : 4c.)

the first division of meiosis, results in the formation of a Holliday junction between markers A and B, which then spreads by branch migration past marker B and is finally terminated between markers B and C, resulting in a recombinant configuration. The result is the formation of two stretches of heteroduplex DNA, encoding information for B on one single strand and b on the other, on each of the chromosomes involved in the recombination event. What can happen to this heteroduplex DNA? The choices are many. One, both, or neither of the heteroduplexes can be repaired. If repair occurs, one strand will be repaired to encode the information

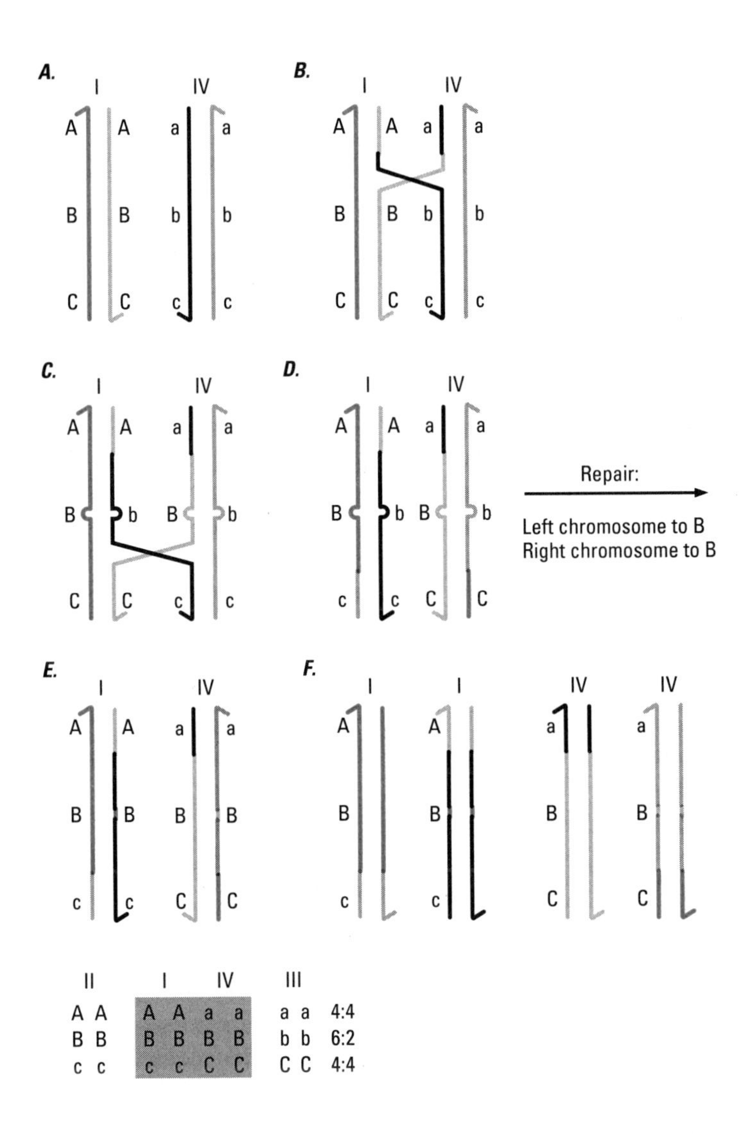

encoded on the other; thus, the choice is twofold: B → b or b → B. Repair involving both heteroduplexes will result in some cases in a 6 : 2 or a 2 : 6 segregation of marker B; in other cases an aberrant 4 : 4 segregation will be observed (see Fig. 4.8). Repair involving only one heteroduplex while

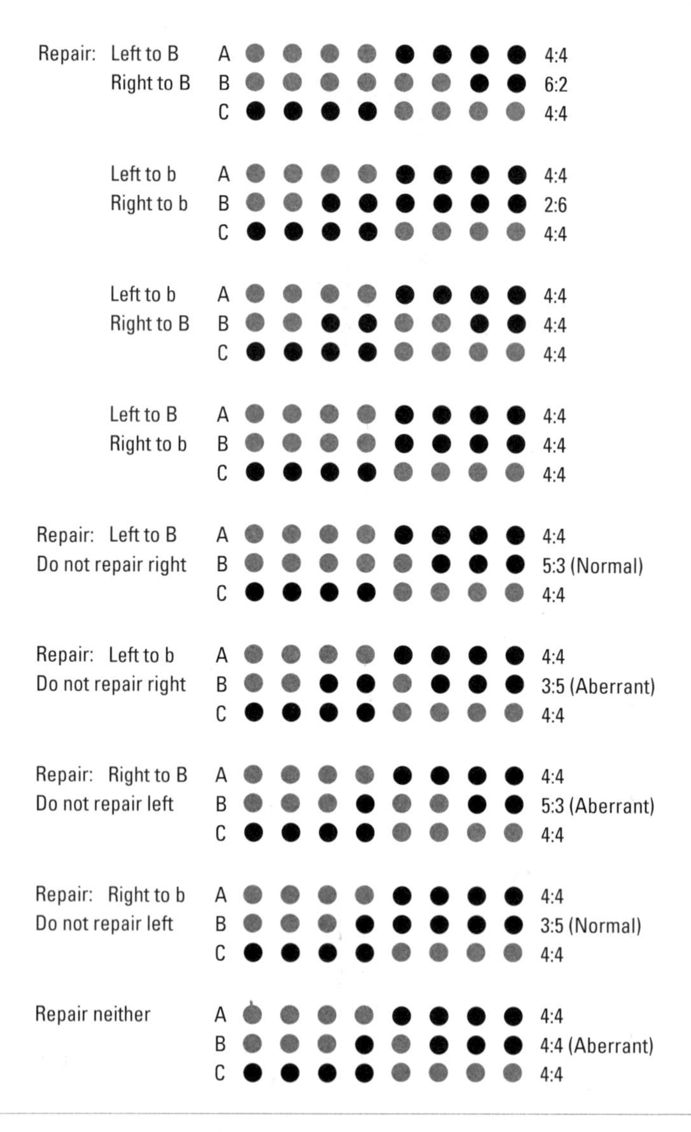

FIGURE 4.8 Examples of asci resulting from gene conversion. Various outcomes of the recombination event depicted in Figure 4.7 are illustrated. The fate of each heteroduplex chromosome during attempted repair is indicated on the left. The order of spores in the asci is the same as in Figure 4.7*F*. The overall segregation of genetic markers is indicated at right. All possible permutations of repaired and nonrepaired heteroduplexes are shown.

the other remains unrepaired will result in a variety of 5 : 3 and 3 : 5 segregant configurations. Neither heteroduplex being repaired can result in a variety of aberrant 4 : 4 segregation patterns.

It should be kept in mind that the Holliday model is not simply an esoteric exercise in cartoons drawn on paper. Aberrant genetic segregation has been observed in a variety of organisms for many decades. Gene conversion was thus a real phenomenon begging for an explanation, and the Holiday model was the first such explanation that made good sense in molecular terms and thus came to be widely accepted.

Unusual Ratios of Gene Conversion Events Point to the Existence of Asymmetric Heteroduplex DNA

The lower fungi are a wonderful model system. Sophisticated genetic experiments can be designed to address very specific questions. It is possible to analyze, for example, the segregation of several closely spaced point mutations within a single gene. Outside genes can be included in the crosses to determine whether recombination was terminated in recombinant or parental configurations.

Studies of this kind produced some interesting observations. First, recombination appears to initiate preferentially at specific sites. Such sites are spaced relatively frequently along the chromosome, on the order of every few thousand base pairs. Although these sites have not yet been analyzed in molecular detail, they are reminiscent of the Chi site in *E. coli.* Second, the frequency of gene conversions is relatively high near such initiation sites and falls off progressively with distance. Third, in any single cross, gene conversion of a marker raises substantially the probability of gene conversion of a neighboring marker. This phenomenon is often referred to as coconversion. In other words, gene conversions seem to occur in clusters. Fourth, reciprocal crossovers of flanking genes are frequently associated with gene conversion.

The detailed examination of coconversion segregation data revealed features that were not easily explained by the Holliday model. In brief, the relative ratios of the various types of gene conversion asci were often found to be significantly different from the predicted values. For example, assuming that repair is random, 50 percent of 5 : 3 asci should be in the "normal" configuration and 50 percent in the "aberrant" configuration (see Fig. 4.8), but in fact normal 5 : 3 asci are often much more prevalent. In some cases, 6 : 2 asci were found to be infrequent for markers close to an initiation point, but their frequency increased for more distal markers (the predicted ratio is 25 percent of all gene conversion events; see Fig. 4.8). Likewise, aberrant 4 : 4 segregation is rare near the initiation point but increases further off.

Unfortunately, it is beyond the scope of the present discussion to analyze gene conversion data in greater detail. Nevertheless, the evidence points to the existence of **asymmetric heteroduplex DNA.** In other words, a lot of the genetic data can be explained if a heteroduplex region was formed on only one of the DNA duplexes involved in the recombination event.

The Meselson-Radding Model

A model was proposed in 1975 by Matthew Meselson and Charles Radding that incorporated asymmetric heteroduplex DNA as one of its hallmark features (see Fig. 4.9). In brief, the model envisions the initiation event as a single nick on one of the partners. The nick is converted to a recombinogenic substrate, such as a single-stranded tail or gap, which invades the opposite DNA duplex. Strand invasion leads to strand exchange catalyzed by a recombinase protein, which results in the formation of a stretch of asymmetric heteroduplex DNA. Through a series of isomerizations a Holliday junction is created (see Fig. 4.10). Once established, the Holliday junction can branch migrate, resulting in a stretch of symmetric heteroduplex DNA. Finally, the Holliday junction is resolved by nicking the appropriate single strands.

The Meselson–Radding model retains some important features of the Holliday model and builds upon those foundations by introducing the concepts of a single initiation event and asymmetric heteroduplex DNA. Doing away with symmetric initiation events at homologous positions on both recombining partners removed a major theoretical difficulty with the Holliday model: In order to initiate recombination, how does the cellular machinery know, without a priori base pairing interaction, where homologous sites are located? In hindsight, asymmetric initiation seems rational and natural. It must be kept in mind, however, that in 1975 the RecA protein, with its fascinating recombinase activities, was still waiting to be discovered. The line of reasoning actually proceeded from the opposite direction. The necessity for asymmetric heteroduplex DNA was more or less dictated by the accumulating genetic data, and the most logical way to postulate its formation was to invoke asymmetric initiation events. It must be pointed out that to this day, the Meselson–Radding model, with some minor modifications, can account for virtually all known data from genetic crosses.

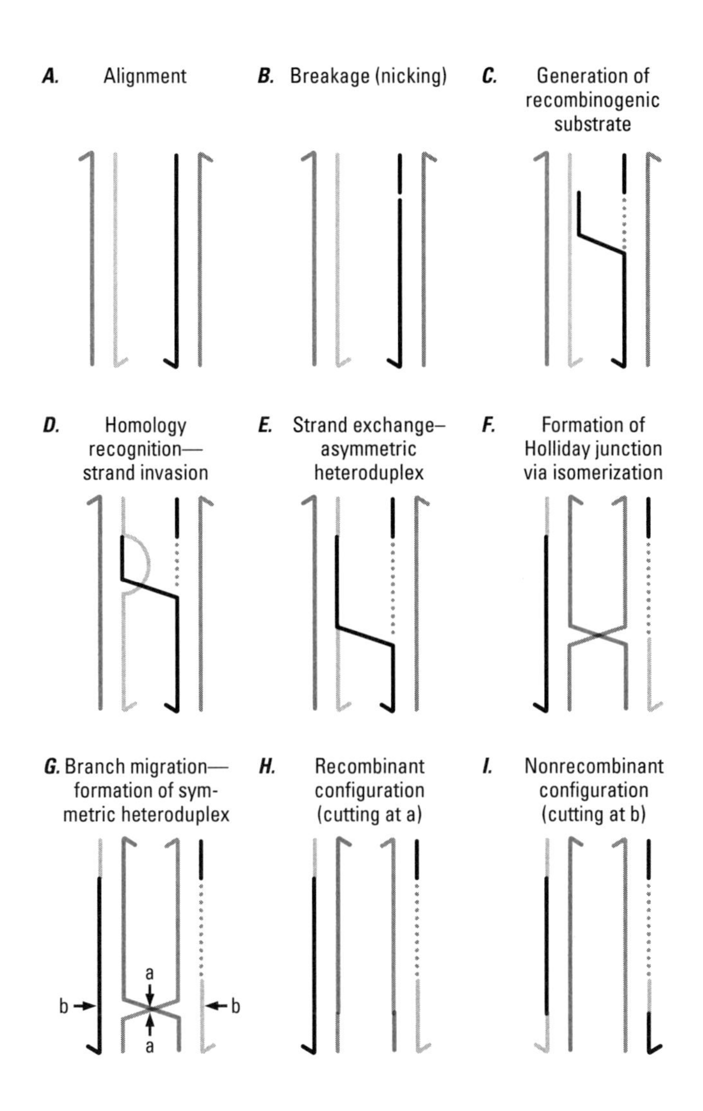

FIGURE 4.9 Meselson–Radding model. The color scheme and layout are the same as in Figure 4.4. (*A*) Chromosomes aligned just prior to recombination. (*B*) A single nick is introduced in one of the chromosomes. (*C*) The nick is converted to a recombinogenic substrate, a single-stranded tail with a 5′ end (the dotted line indicates new DNA synthesis). (*D*) The tail in the process of homology comparison with the opposite chromosome. (*E*) Strand exchange has been initiated following the successful recognition of homology; this results in the formation of a region of asymmetric heteroduplex. (*F*) A Holliday junction is formed through an isomerization of the structure in *E*. The isomerization involves several steps, shown in detail in Figure 4.10. (*G*) Branch migration of the resultant Holliday junction, leading to formation of two regions of asymmetric heteroduplex. (*H*) Resolution of Holliday junction resulting in a reciprocal crossover event. This is achieved by introducing two nicks at positions marked a in *G*. (*I*) Resolution of Holliday junction not resulting in a reciprocal crossover event. This is achieved by introducing two nicks at positions marked b in *G*.

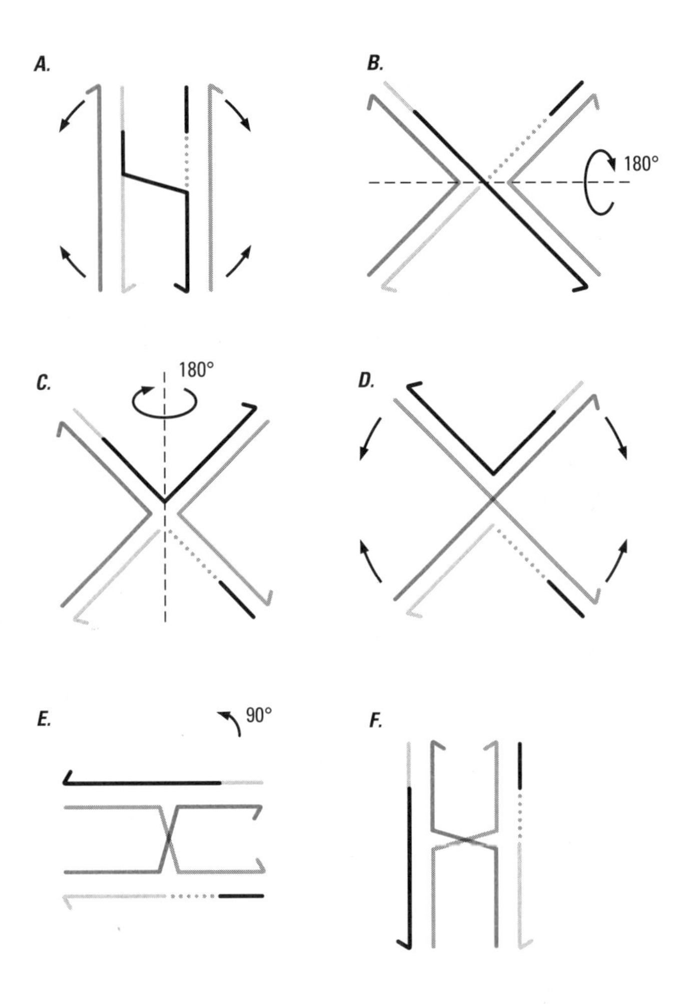

FIGURE 4.10 Creation of a Holliday junction by isomerization of an asymmetric strand exchange event. (*A*) (same as Figure 4.9*E*) Product of an asymmetric strand invasion and exchange event. The first operation is to draw the arms at the top and bottom of the figure outwards (see arrows) to transform the side-by-side configuration into a figure X configuration. (*B*) The next operation is to rotate the rightward arms 180° around a horizontal axis running through the center of the junction. The axis of rotation is indicated by a dotted line and the direction of rotation by an arrow. (*C*) The next operation is to rotate the upward arms 180° around a vertical axis running through the center of the junction. Again, the axis of rotation is indicated by a dotted line and the direction of rotation by an arrow. (*D*) The next operation is to bring the arms together (see arrows) to transform the figure X configuration into a side-by-side configuration. (*E*) The last operation is a simple 90° rotation to put the chromosomes in their starting orientation. The axis of rotation, indicated by a dot, is oriented in and out of the plane of paper; the arrow shows the direction of rotation. (*F*) This results in a structure identical to the structure in Figure 4.9*F*.

Repair of Double-Strand Breaks During Mitotic Recombination in Yeast Leads to a New Model of Recombination

In the late 1970s it became possible to introduce exogenous DNA into yeast cells by a process called **transfection.** The DNAs used in those experiments were small, circular, double-stranded molecules referred to as **plasmids.** For technical reasons, transfections can be performed only with vegetatively growing cells, either haploid or diploid (Fig. 4.1). If the exogenous DNA molecules contained an origin of DNA replication, the frequency of transfection was found to be relatively high. Under such conditions, the frequency of transfection reflects only the efficiency with which the DNA penetrates into the cell. Once inside, the DNA molecules are continuously replicated and relatively efficiently partitioned to daughter cells.

If the exogenous molecules lacked an origin of DNA replication, the frequency of transfection precipitously decreased. Nevertheless, stably transfected cells could be recovered, but only under one condition: The exogenous DNA had to contain sequences homologous to some part of the resident chromosomes. Examination of the fate of the transfected DNA revealed that in all cases it was integrated at the homologous chromosomal locus by a process indistinguishable from homologous recombination (see Fig. 4.11*A*). This was not altogether surprising, since the existence of homologous recombination activity in mitotically dividing cells had been known from previous studies. The frequency of homologous recombination in mitotic cells is, however, much lower than during meiosis. The low transfection efficiency of origin-lacking DNAs is therefore due to the compounded effects of transfection and recombination frequencies.

Manipulation of the exogenous DNA *in vitro* prior to its transfection into yeast cells led to some surprising observations. First, introduction of a double-strand break in the region of homology greatly increased the recovery of stably transfected cells. The products of the recombination event were indistinguishable from experiments in which a circular molecule was used (see Fig. 4.11*B*). Double-strand breaks in parts of the molecule outside the region of homology did not produce this effect. The striking increase in transfection efficiency was due to an increase in the frequency of homologous recombination, since linear DNA molecules do not penetrate into cells any better than circular ones. In fact, double-stranded ends appeared to be acting as efficient recombinogenic substrates. An even more unexpected result was obtained when instead of a double-stranded break, an actual gap was introduced into the region of homology. Both the level of stimulation of homologous recombination

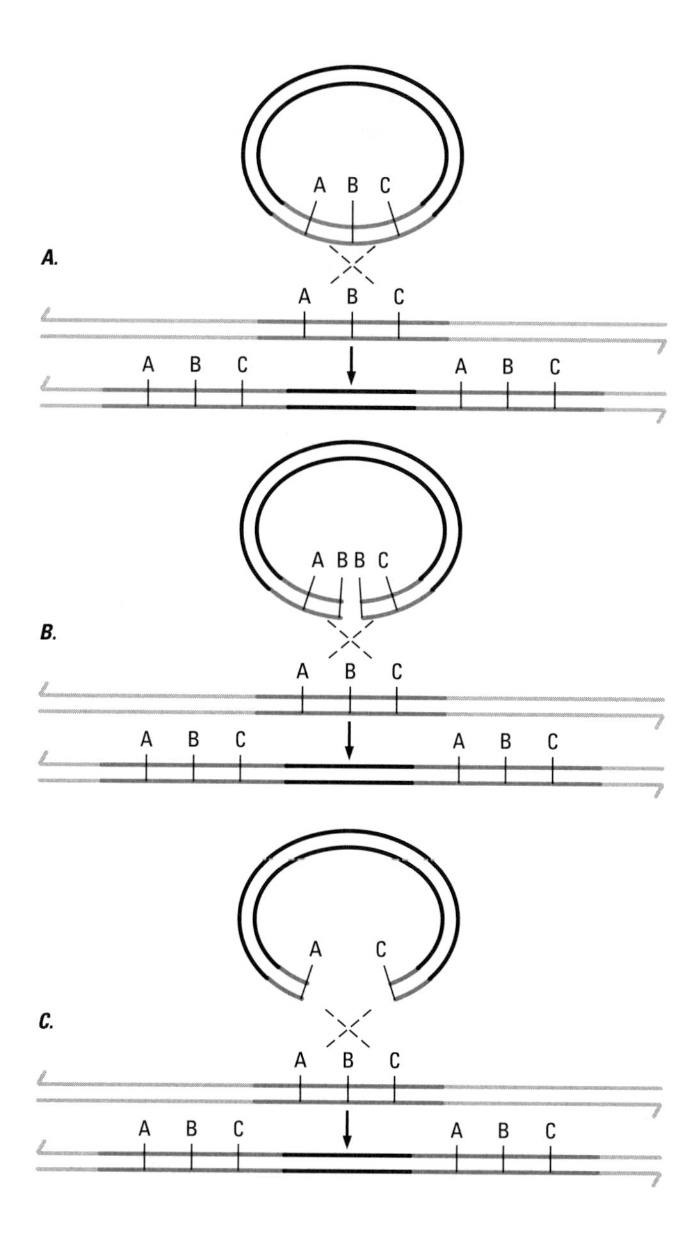

and the configuration of the observed products remained the same (see Fig. 4.11*C*). In other words, information was transferred unidirectionally from the chromosome to repair the gap in the plasmid. Both of these observations seemed paradoxical because neither can be accounted for in a straightforward fashion by the Meselson–Radding model.

FIGURE 4.11 (*Left*) Recombination of intact and gapped yeast plasmids with homologous chromosomal sequences. The chromosome is shown as a linear DNA molecule (green). The circular plasmid, shown in black, contains a region of homology with the chromosome highlighted in green. Three specific sites in the regions of homology, designated as A, B, and C, are included as reference points. The plasmid sequences (light green) are not homologous to any part of the chromosome. (*A*) Recombination event between a circular plasmid molecule and the chromosome. The crossover in the region of homology is indicated with dotted lines. The recombination event can be modeled using the Meselson–Radding paradigm as a single reciprocal crossover; the outcome is shown at the bottom of the drawing. (*B*) The plasmid molecule has been interrupted with a double-stranded break at site B. No sequences have actually been removed from the region of homology. Molecular analysis of recombinant products generated in an experiment such as this revealed their structures to be identical to that shown in part *A*. (*C*) The plasmid molecule has been interrupted with a double-stranded gap by cutting it at two sites, A and C. The sequences between sites A and C have thus been removed from the region of homology. Again, molecular analysis of recombinant products generated in an experiment such as this revealed their structures to be identical to that shown in part *A*.

The Double-Strand-Break Repair Model of Recombination

In 1983, Jack Szostak, Terry Orr-Weaver, Rodney Rothstein, and Frank Stahl presented a model that attempted to reconcile these observations in molecular terms (see Fig. 4.12). In brief, the model envisions the initiation event as a double-strand break on one of the partners. The break is converted into two recombinogenic single-stranded tails, one of which invades the opposite DNA duplex. A small D loop is created and subsequently enlarged by repair synthesis primed from the invading single-stranded tail. The D loop eventually extends into sequences complementary to the single-stranded tail on the other side of the gap, and the two partners can anneal. A second round of repair synthesis is then initiated. The net result is repair of the gap by two successive rounds of single-strand repair synthesis and formation of two regions of asymmetric heteroduplex bracketed by two Holliday junctions. Once established, the Holliday junctions can branch migrate, resulting in further stretches of symmetric heteroduplex DNA. Finally, the Holliday junctions are resolved by nicking the appropriate single strands.

Although the initial impetus for the model was a desire to explain double-strand-gap repair in mitotically growing cells, it was soon realized that it could easily be extended to explain most observed features of meiotic recombination and gene conversion. The major points of depar-

A. Alignment

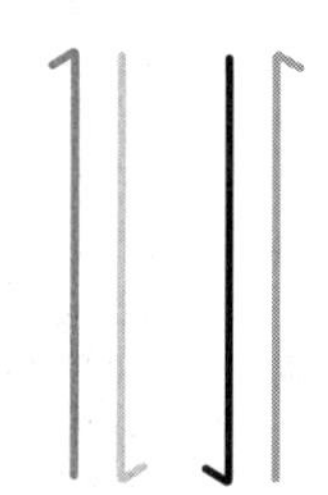

B. Introduction of a double–strand break

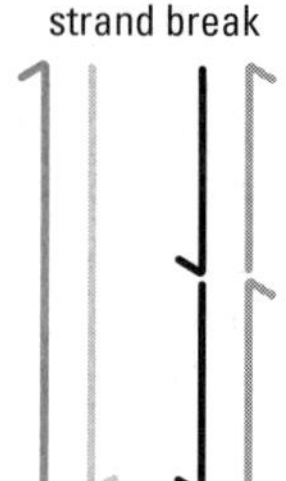

C. Generation of recombinogenic subtrates

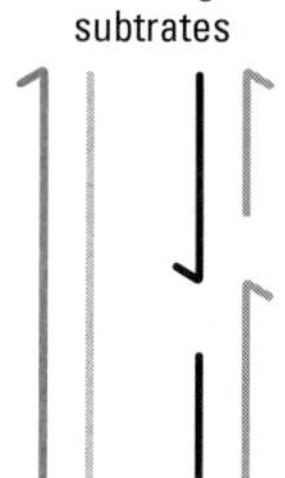

D. First homology recognition—strand invasion

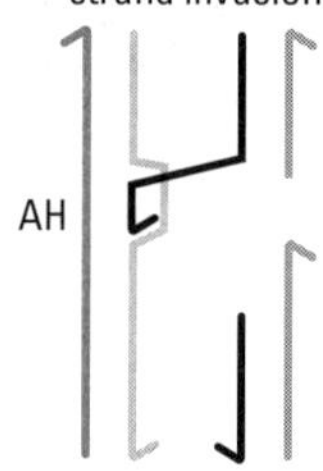

E. Second homology recognition

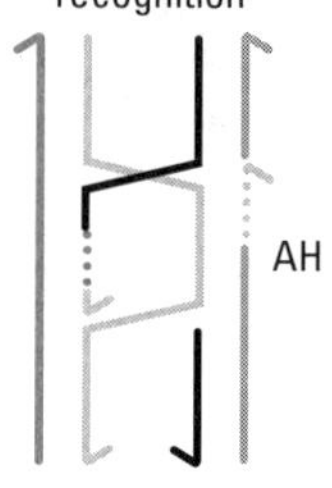

F. Formation of two Holliday junctions

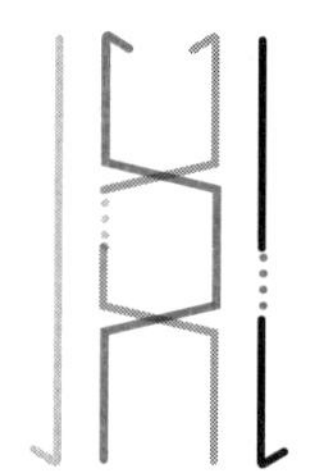

G. Branch migration

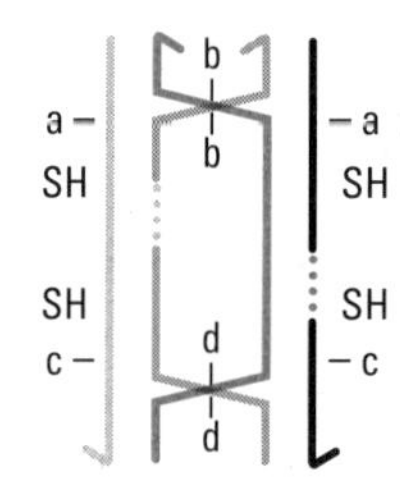

H. Recombinant configuration (cutting at a and d)

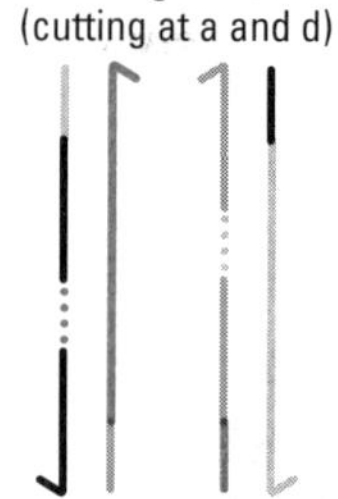

I. Recombinant configuration (cutting at b and c)

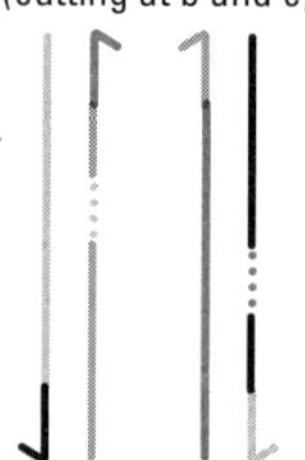

J. Nonrecombinant configuration (cutting at a and c)

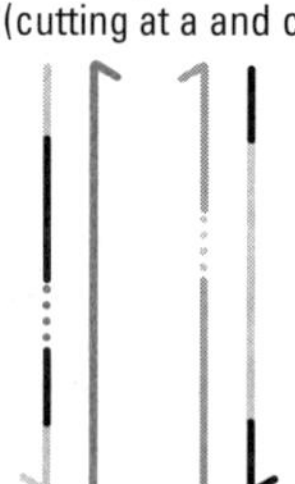

K. Nonrecombinant configuration (cutting at b and d)

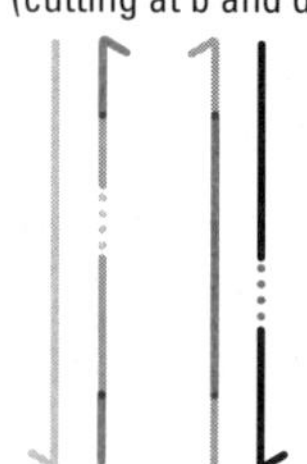

FIGURE 4.12 (*Left*) Double-strand-break repair model. The color scheme and layout are the same as in Figures 4.4 and 4.9. (*A*) Chromosomes aligned just prior to recombination. (*B*) A single double-stranded break is introduced into one of the chromosomes. (*C*) The break is shown converted into two recombinogenic substrates by the action of a 5′→3′ double-strand specific exonuclease. The result is the production of single-stranded tails with 3′ ends. (*D*) One of the tails is in the process of homology comparison with the opposite chromosome. Following the successful recognition of homology a D loop is displaced; this results in the formation of an asymmetric heteroduplex track (labeled AH). (*E*) The 3′ end of invading strand is acting as a primer for DNA synthesis (dotted line), thus displacing an ever larger D loop. The D loop is enlarged until it can base pair with the second single-stranded tail, creating a second asymmetric heteroduplex region (labeled AH). When this happens, the 3′ end of the tail can act as a primer for a second round of DNA synthesis. (*F*) Two Holliday junctions formed through an isomerization event (shown in detail in Fig. 4.10). (*G*) Branch migration of the two Holliday junctions, leading to the formation of regions of symmetric heteroduplex (labeled SH). (*H, I*) Possible resolutions of the Holliday junctions resulting in a reciprocal crossover event. This is achieved by introducing nicks at the indicated positions. (*J, K*) Possible resolutions of the Holliday junctions not resulting in a reciprocal crossover event. This is achieved by introducing nicks at the indicated positions.

ture of the model are, first, initiation of recombination by a double-strand break and, second, the involvement of two Holliday junctions in a single recombination event. Gene conversion is envisioned to occur both by gap repair as well as mismatch repair of the asymmetric and symmetric heteroduplex DNA tracts flanking the gap.

Multiple Pathways of Homologous Recombination Can Be Active at the Same Time

Which model is correct? The answer may well turn out to be both. Unfortunately, recombination has proved to be exceptionally recalcitrant to biochemical analysis. To this day, a recombination reaction cannot be performed in the test tube in its entirety, from initiation to resolution. Only certain components, such as, for example, strand invasion or resolution of Holliday junctions can be performed in vitro. To distinguish between specific mechanistic details, however, a detailed biochemical analysis is absolutely essential. The problem is that the large body of genetic data simply lacks the required fine focus; in most cases either model can be used to explain a particular observation by invoking certain additional assumptions. The argument thus often boils down to which model requires fewer additional assumptions or which assumptions appear less ad hoc.

The notion of multiple recombination pathways initially came from genetic studies. In *E. coli,* for example, numerous recombination-deficient mutants have been isolated over the years. Not all mutants, however, block recombination between all substrates; for example, recombination during bacterial conjugation (mating) is blocked by certain mutants that have little or no effect on recombination between plasmids. During conjugation, the donor incoming DNA is linear, allowing the RecBCD enzyme to enter and generate recombinogenic substrates. In contrast, recombination between two circular molecules, such as plasmids, depends on different enzymes for the generation of recombinogenic substrates. Such studies led to the notion of recombination pathways, somewhat analogous to metabolic pathways, in which multiple enzymes act in succession to transform substrates into products. The picture of recombination pathways is clouded somewhat by the fact that some activities, such as the RecA protein or DNA ligase, are required by all pathways. In addition, under certain circumstances enzymes can substitute for each other between pathways. Nevertheless, it is clear that *E. coli* cells can recombine DNAs in more than one strictly defined way.

The picture of multiple pathways is also appropriate when higher organisms are considered. In yeast, for example, it has been clear for a long time that recombination during meiosis and recombination during mitosis are by and large two distinct processes. During meiosis, recombination is proficient and its function is sexual reproduction; in contrast, during mitosis, recombination activity is much lower and its function is postreplication repair of DNA. Some yeast mutants have been found that affect both meiotic and mitotic recombination, but most seem to affect either one or the other process. In addition, there are indications that mitotic recombination may be in fact a combination of several overlapping pathways, somewhat analogous to the situation found in *E. coli.*

The models presented in this chapter should thus be viewed as being flexible processes rather than rigid step-by-step pathways. When cells go about their business, they may use a few steps from one model, a few from another when convenient, and so forth. The current information concerning the RecBCD pathway in *E. coli,* for example, is most consistent with the Meselson–Radding model, whereas gene conversion in yeast seems to be more in line with the Szostak et al. model. Most important, much new information about the detailed mechanics of recombination remains to be discovered, and it is certain that nature still holds many surprises for us.

Suggested Readings

Holliday, R. (1968). Genetic recombination in fungi. In *Replication and recombination of genetic material,* J. W. Peacock and R. D. Brook, eds. Canberra: Australian Academy of Science, pp. 157–174.

Meselson, M. S., and Radding, C. M. (1975). A general model for genetic recombination. *Proc. Natl, Acad. Sci. USA* 73: 358–361.

Stahl, F. W. (1979). *Genetic recombination: Thinking about it in phage and fungi.* San Francisco: W. H. Freeman.

Szostak, J. W., Orr-Weaver, T. L., Rothstein, R. J., and Stahl, F. W. (1983). The double-strand break repair model for recombination. *Cell* 33: 25–35.

5

GENE TARGETING

What is gene targeting? Gene targeting is, quite simply, the willful modification of genetic information in a living organism. Three criteria can be used as a guideline. First, the process must be **directed,** affecting only the locus of choice. Second, the process must be **specific,** such that a predetermined sequence can be inserted or substituted at the target locus. Finally, to be of practical significance, the process should be **efficient.** These points are useful because they distinguish gene targeting from other ways of manipulating genomes.

Why is gene targeting useful? Gene targeting allows the direct isolation of mutant organisms. In the test tube, the possibilities for manipulating DNA molecules are just about infinite. DNA molecules can be sequenced, they can be cut at predetermined positions, the fragments can be rearranged, and pieces with completely novel sequences can even be synthesized from scratch. The crux of the problem is to take such in-vitro-manipulated DNA molecules and incorporate them, using the above-mentioned three criteria, into the genomes of living cells.

Why do we want to isolate mutant organisms? In the context of this discussion it should first be reemphasized that the term mutation does not imply anything about the effect on the organism, which could be detrimental, neutral, or even beneficial. In other words, **mutant** simply means different from what existed before. The use of mutant organisms in research goes back to the very beginnings of biology. In modern molecular biology, deciphering gene action has become the central focus: What are the functions of particular genes in the life-cycle of an organism, and when and how do they do it? Invaluable insights into the function of a gene can be gained through the study of mutations that affect its normal activity; in other words, by analyzing the abnormal, one can learn about the function of the normal.

Most important, such analysis often allows the establishment of cause-and-effect relationships that are to a large degree the cornerstone of experimental biology. This is usually not possible by simply observing the normal behavior of organisms. For example, it may be suspected that a certain physiological function is essential for a particular behavior; the isolation of a mutant organism defective in that function that cannot perform the specific behavior can in many cases be used as rigorous proof of the hypothesis. A significant fraction of the landscape of molecular biology as we know it today has been elucidated by the combined attack of genetic manipulations coupled with the biochemical analyses of the consequences of the mutational alterations.

The practical applications of the isolation of mutant organisms are also far reaching. For example, it may be possible to isolate mutants that possess enhanced beneficial traits, decreased detrimental traits, completely novel or unique traits, and so forth. There is really nothing new in our desire to perform such feats; husbandry and selective breeding of agricultural animals and crops is an ancient practice whose aim is gradual stock improvement. The only major point of departure is that modern genetic engineers have much more direct and efficient tools to get the task done. In terms of human medicine, one obvious application is the correction of inborn errors of metabolism. The issues surrounding genetic engineering and gene therapy will be more thoroughly explored in the epilogue.

If the definition of gene targeting given at the beginning of this chapter is carefully considered, it becomes evident that the only way to achieve the task is to take advantage of homologous recombination. Genetic engineers know how to manipulate DNA in test tubes, but to manipulate it directly inside living cells and have them survive the operation is not possible with current technology. The first key to gene targeting is to present the exogenous DNA to cells in a sufficiently appetizing fashion to trick them into mistaking it for their own. In effect, the exogenous DNA falls into the cogs of the homologous recombination machinery that is continuously churning away in all living cells. The second key to gene targeting is to efficiently recover the cells that have successfully executed the desired homologous recombination event.

Mammalian Cells Propagated in Culture Are in Many Ways Similar to Unicellular Microorganisms

Modern molecular genetics is built on the foundations of classical genetic studies in microorganisms, particularly bacteria such as *Escherichia coli* and its viruses, the yeast *Saccharomyces cerevisiae*, and some higher fungi. The principal advantages of these model systems is that they employ small unicellular organisms, with simple life cycles, basic nutritional requirements, and very rapid growth rates. In practical terms this translates to being able to perform many experiments in a small space using cheap growth medium. In the laboratory microorganisms such as *E. coli* or *S. cerevisiae* can be propagated either in a liquid medium or on the surface of a semisolid medium (see Fig. 5.1*A*). The principal utility of liquid medium is the production of large numbers of organisms, whereas semisolid medium is used to propagate isolated, genetically homogeneous populations. When a single cell is placed on the surface of a semisolid medium and starts to divide, the progeny accumulate in that location and produce a small clump referred to as a **colony** (see Fig. 5.1*B*). This is possible if the cells are either not motile or cannot move very far on the surface of the medium. Since all the cells in a colony are derived from a single ancestor cell by asexual reproduction, the population is genetically homogeneous and is often referred to as a clonal population, or **clone.**

Under appropriate conditions, cells of higher organisms can be explanted from the body and grown indefinitely in culture. For historical reasons the procedures used for propagating mammalian cells are collectively referred to as **tissue culture,** although today the term cell culture would be more appropriate. In comparison to *E. coli* and *S. cerevisiae,* the nutritional requirements of mammalian cells are very complex. Indeed, the formulation of appropriate culture media took many years. Nevertheless, in many ways tissue culture cells behave as large, slow-growing microorganisms. The nutritional requirements of most cell types by now have been adequately worked out, and numerous good medium formulations are readily available. Depending on the tissue from which the cells were derived, mammalian cells can be propagated in liquid culture, grown on solid surfaces, or both. When grown on solid surfaces, under the right conditions clonal colonies can be formed.

Research scientists are always asking, "How many experiments can be performed in a given amount of time?" Quite simply, experiments that consume too many resources or take too long to execute are usually not attempted. From this practical, utilitarian perspective, the two major differences between tissue culture cells and the favorite research microorganisms *E. coli* and *S. cerevisiae* are **growth rate** and **cell size.** It is useful to consider just how crucial these parameters are in terms of designing experiments.

E. coli and *S. cerevisiae* under optimal conditions can rapidly grow to high densities. *E. coli* cells can divide every 20 to 30 minutes; this growth parameter is often referred to as the generation time, or **doubling time.** *S. cerevisiae* is somewhat slower, with a doubling time of 50 to 60 minutes.

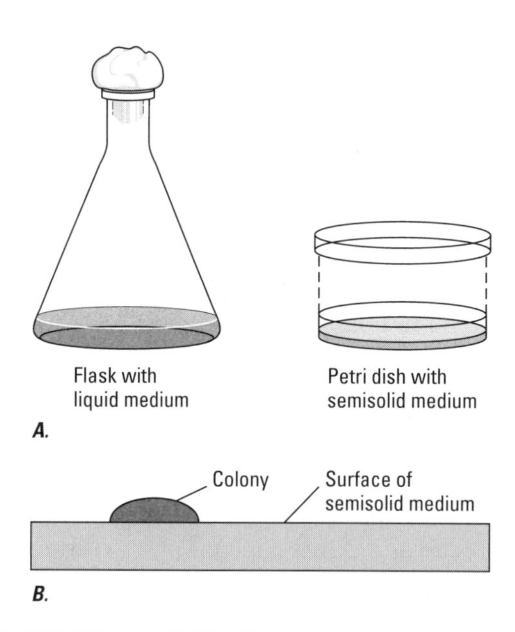

FIGURE 5.1 Growth and propagation of microorganisms in the laboratory. (*A*) Use of two principal types of media, liquid and semisolid. Liquid medium is simply an aqueous solution of the required nutrients. An appropriate culture vessel is inoculated with cells, which then divide until the availability of nutrients and/or the accumulation of waste products in the medium become limiting. At that point the culture reaches its saturation point, sometimes also referred to as stationary phase. Semisolid medium is liquid medium with the addition of agar, which causes the medium to congeal to a consistency resembling that of firm jello. (*B*) Close-up, in a side-view cross section, of a colony growing on the surface of semisolid medium.

In contrast, mammalian cells have minimum doubling times of 12 to 14 hours; many cell types require upwards of 24 hours. These differences become dramatic when it is considered how long it takes a single cell to grow into a colony. Under optimal conditions, *E. coli* will form colonies overnight, *S. cerevisiae* in two days, and mammalian cells in approximately 10–14 days. In terms of culture density, saturated liquid cultures of *E. coli* and *S. cerevisiae* contain approximately 2 billion (2×10^9) cells and 100 million (1×10^8) cells per milliliter, respectively, whereas mammalian cells will grow to densities of only about 1 million (1×10^6) cells per milliliter.

Mammalian cells are real giants next to *E. coli* and *S. cerevisiae* (see Fig. 5.2*A*). An *E. coli* cell is rodlike in shape, measuring 1 by 3 micrometers; a yeast cell is spherical with a diameter of 6 to 7 micrometers. In contrast,

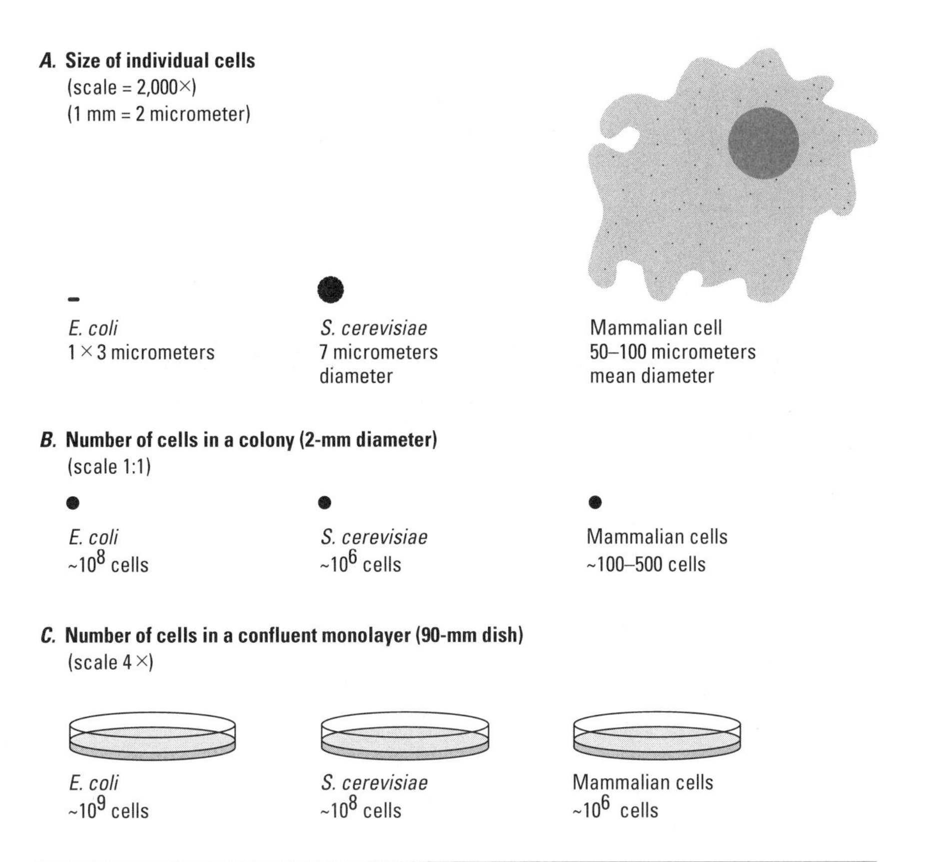

FIGURE 5.2 Relative sizes of *E. coli, S. cerevisiae,* and mammalian cells. (*A*) Cells magnified one thousandfold. (*B*) Typical colonies are life-size; the approximate number of cells per colony is also indicated. (*C*) Classic Petri dishes one fourth their true size; the average number of cells that can be comfortably accommodated per dish is also indicated.

mammalian cells are highly variable in shape, often assuming very irregular contours. Cells grown on a solid surface typically assume flattened shapes, somewhat like pancakes, measuring 50 to 100 micrometers in diameter. The consequences of these size differences are also dramatic. A single colony of about 2 millimeters in diameter contains up to a hundred million (10^8) *E. coli* cells, a few million (10^6) *S. cerevisiae* cells, but only several hundred (10^2) mammalian cells (see Fig. 5.2*B*). Cell size also directly determines how many cells can be accommodated per unit surface area. The classic Petri dish, for example, has a diameter of approximately 9 centimeters. If cells are spread into a uniform, tightly packed layer but not piled on top of each other, a single Petri dish will accommodate approximately 2×10^9 *E. coli* cells, 1×10^8 *S. cerevisiae* cells, but only 2×10^6 mammalian cells (see Fig. 5.2*C*). This means that one thousand Petri dishes are needed to hold as many mammalian cells as one dish of *E. coli!*

A Primer on Some Classical Methods and Jargon of Microbial Genetics

All genetic manipulations involve one basic principle: **selection.** In other words, the experimenter gets to choose which of his experimental organisms are allowed to live and propagate and which are left by the wayside. The principle is the same whether one considers a prehistoric farmer who always chose, from season to season, the seeds from his tallest corn plants for next year's planting or the strongest calf for further breeding or a microbial geneticist examining thousands of individual colonies in his Petri dishes.

Perhaps the most basic concept of genetics is the notion of a **genetic marker.** A marker is, quite simply, a trait that can be distinguished by one means or another in the experimental organism. Most important, the trait must be genetic in nature; in other words, it must show a stable and reproducible pattern of inheritance. The seed color of peas, for example, can be either yellow or green. Since these are stably inherited traits, they can be regarded as genetic markers. Similarly, corn plants can come in two varieties, tall and short. Although environmental stress—for example, drought—can make all corn plants short, plants bearing the dwarf marker will always be shorter, in the same field, than their nondwarf cousins.

Today, of course, we know that genetic markers are, in molecular terms, genes. One example that can be used to illustrate the case is carbohydrate metabolism in *E. coli.* Several simple sugars, such as glucose, maltose, fructose, lactose, and so forth, can be utilized by the cells as the sole source of energy as well as precursors for the synthesis of macromolecules. *E. coli* cells will grow quite happily on a medium containing, for example, maltose as the only organic nutrient. It is now known that the assimilation and further metabolism of maltose is catalyzed by several enzymes acting in a pathway. Each of these enzymes is encoded by a gene in the *E. coli* genome. A mutation in one of the genes rendering its product—for example, the maltose transport protein—nonfunctional would eliminate the ability of that cell to utilize maltose. In the genetic jargon, such cells would be referred to as **maltose negative,** or Mal$^-$, since they cannot grow on maltose as their primary nutrient. Most important, since the cause of maltose negativity is a change in the DNA encoding the relevant gene, all the progeny of such a cell would likewise be maltose negative. Such cells could be, however, easily propagated on a medium supplemented with another sugar.

Another basic concept of genetics is the notion of **genotype and phenotype.** The genotype describes the status of the genome; the consequences of the genotype, such as visible changes in the appearance or behavior of the organism, are called the phenotype. Maltose metabolism can again be used to illustrate the distinction. Seven genes are required for maltose assimilation in *E. coli;* they are designated simply *malE, malF, malG, malK, malQ, malT,* and *lamB* for identification purposes. The maltose transport protein happens to be the product of the *malF* gene. A clone of *E. coli* cells carrying a loss-of-function mutation in, for example, the *malF* gene would be described as possessing a *malF*$^-$ genotype. The result of this genotype would be a maltose negative, or Mal$^-$, phenotype. It is evident that the two are nonequivalent, since a Mal$^-$ phenotype can be caused by several different genotypes, for example, malE$^-$ or malG$^-$.

Any differential in an ability to grow on a defined medium can be exploited as the basis for a selection. To put it simply, Mal$^-$ cells cannot form colonies on maltose medium, whereas Mal$^+$ cells can. If, for example, a million Mal$^-$ cells and five Mal$^+$ cells are spread on a maltose plate, on the next morning five colonies will be observed. A selection scheme can thus be developed in which the *malF* gene, which encodes one of the enzymes involved in maltose transport, can act as a **positive marker.** In other words, the presence of the *malF* gene can be positively selected in *malF*$^-$ cells.

A simple manipulation of the growth medium can, however, turn the whole picture around. A medium can be prepared that contains another sugar as the nutrient and, in addition, a toxic analog of maltose. If the toxic analog is transported into the cell, it will kill it. Under these conditions, Mal$^-$ cells can form colonies, because they will not assimilate the toxic analog but will be allowed to grow on the other sugar as their nutrient. Mal$^+$ cells will, however, take up the analog and die. Thus, in this particular selection scheme, the gene encoding the normal enzyme for maltose transport can act as a **negative marker.**

In the jargon of the microbial geneticist, the term selection has a rather specific meaning. A selection is a procedure in which only those cells that satisfy the conditions of the selective medium are capable of growth. Another approach is to formulate a medium that allows all cells to develop into colonies but provides, in addition, some visible characteristic by which clones bearing a desired marker can be distinguished. Such a procedure is usually referred to as a **screen.** For example, an **indicator medium** can be developed that contains maltose, several other nutrients, and special compounds that change color when the maltose is utilized by the cells. Thus, both Mal$^+$ and Mal$^-$ cells are able to form colonies, but all Mal$^-$ colonies have a pale, white appearance, whereas Mal$^+$ colonies appear colored, for example, deep red. The composition of indicator media can be complex and is beyond the scope of the discussion at hand. A simple way to explain the basic distinction between a selection and a screen is that in a selection everyone dies except those that pass the test, whereas in a screen everyone gets to live, but the desired colonies look different. Depending on the particular experimental goal, a selection or a screen may be more appropriate; the capability of choosing one or the other provides the experimenter with an advantageous degree of flexibility.

Genetic selections and screens are indirect. This stems from the simple fact that it is impossible to directly change a desired DNA sequence in the genome of a living cell; in other words, a selective medium can only select for a particular phenotype. All good geneticists are thus always asking themselves, "Are there any other mutations, in addition to the desired ones, that will satisfy this selection or screen?" Since the success of most experiments hinges on obtaining a very specific result, it is an issue of considerable importance and leads to the concept of **background.** Consider, for example, the design of an experiment to isolate malF$^-$ mutations. A medium with a toxic maltose analog selects for Mal$^-$ cells, but will all Mal$^-$ colonies contain the desired mutations? As indicated earlier, the answer is no, since mutations in other mal genes will also yield Mal$^-$ colonies. In fact, *malF* mutants will be only a small fraction of the total Mal$^-$ colonies. This particular selection thus suffers from a relatively high background, and it may be a good first step in the experiment as long as additional selections or screens are used subsequently to zero in on the desired mutant.

Mammalian Tissue Culture Cells Are Extremely Difficult to Manipulate Genetically

Although mammalian cells grown in tissue culture are in some ways similar to unicellular microorganisms, several factors present severe impediments to the kinds of genetic experiments that can be performed with them. Two reasons, their large size and slow growth rate, have already been explored. The other major impediment is the great difficulty in selecting mutations and the resultant paucity of suitable markers.

In *E. coli,* which is a haploid organism, it is relatively easy to select loss-of-function mutations. Since each cell contains only one copy of, for example, the *malF* maltose transport gene, a single mutation can yield a Mal^- cell. If the analogous mutation were sought in a diploid organism, two independent mutations would have to occur. This is because a cell containing one normal and one mutant maltose transport gene, $malF^-/malF^+$, a condition referred to as **heterozygous,** can still transport maltose. In the genetic jargon, the $malF^-$ mutation is referred to as **recessive,** because its presence can be masked by the presence of a normal copy of the gene. The normal $malF^+$ copy of the gene is referred to as **dominant,** and a $malF^+/malF^+$ or a $malF^-/malF^-$ combination of markers is designated as being **homozygous.** In some organisms, such as *S. cerevisiae,* the problem of isolating recessive mutations can be overcome simply by performing selections during the haploid phase of their life cycle.

In organisms that do not have haploid phases, recessive mutations can be isolated by a two-step process of mutagenesis followed by mating (see Fig. 5.3). This particular scheme allows mutations, masked by their heterozygous condition in the mutagenized individuals, to segregate and appear in future generations. Unfortunately, mammalian cells in culture cannot mate; neither do they possess a haploid phase in their life cycle.

Since most mutations result in the loss of function of a gene, and most loss-of-function mutations are recessive, it is easily appreciated why it is so difficult to select mutations in mammalian cells. This is compounded by the complex nutritional requirements of mammalian cells and the resultant complexity of medium formulations. Only a small number of traits can be conveniently selected in culture. As a result, few markers are available to geneticists working with mammalian tissue culture cells.

Foundations of Gene Targeting in *S. cerevisiae*

The basic procedures of gene targeting were established in the late 1970s and early 1980s in experiments with *S. cerevisiae.* The same ideas were applied during the mid-1980s in experiments using mammalian tissue culture cells, but it was not until late in 1988 that the techniques were perfected and became available as generally applicable genetic tools.

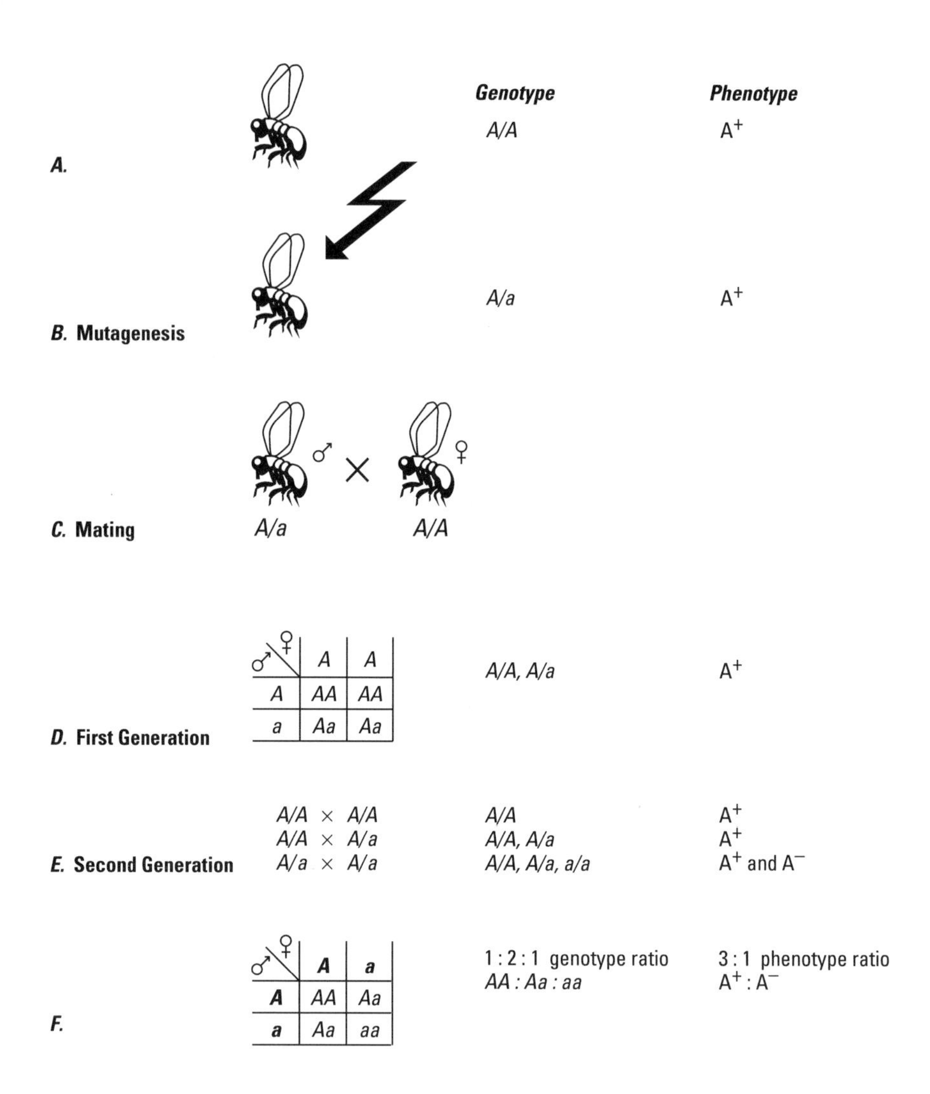

FIGURE 5.3 Isolation of recessive mutations in diploid organisms. The fruit fly *Drosophila melanogaster*, another favorite experimental organism, is used as an example. (*A*) Fly with a hypothetical genotype *A/A* and a corresponding phenotype A^+. (*B*) The fly is mutagenized with, for example, X rays, resulting in a loss-of-function mutation in one of the copies of gene A. The genotype thus becomes *A/a*, but the phenotype remains A^+. (*C*) The mutagenized fly is mated with a normal (*A/A*) fly. (*D*) Expected progeny from this mating: an equal mixture of *A/A* and *A/a* genotypes. The phenotypes of all first-generation progeny are still A^+. The final step of the experiment is to allow the first generation progeny to mate randomly with each other. Three possible matings can take place: *A/A* × *A/A*, *A/A* × *A/a*, and *A/a* × *A/a*. (*E*) Expected genotypes and phenotypes resulting from each of these matings. Only the mating of two heterozygotes can produce *a/a* progeny. (*F*) Expected ratios of genotypes and phenotypes resulting from a second-generation mating of two heterozygotes.

The first breakthrough towards gene targeting was the ability to introduce exogenous DNA into yeast cells by a procedure usually referred to as **transfection.** The ability of a DNA molecule to penetrate into a cell would be of little use, however, if it were not stably inherited in subsequent generations. To assure inheritance, three general criteria must be satisfied: replication, partitioning, and selectability. A DNA molecule possessing all these qualities is often referred to as a **vector,** because it can be used as a general vehicle for introducing other DNA sequences into cells. Most vectors today are man-made DNA molecules constructed for a specific purpose.

To achieve replication, a DNA molecule must possess on **origin of replication.** Origins of replication are specific sequences that are recognized by the replication machinery early in *S* phase and serve as initiation points for the duplication of DNA. Yeast origins of replication are quite short, on the order of 100 base pairs, and are found in chromosomes approximately every few thousand base pairs. In addition to the linear chromosomes, a naturally occurring small circular molecule has been found in many yeast strains. DNA molecules that exist separately from the bulk of the genome are referred to as **episomes;** the yeast episome in question was given the simple descriptive name 2-micron circle (micron is the shortened form of micrometer). A chromosomal origin initiates only a single round of replication during each *S* phase. In contrast, the origin of replication found on the 2-micron circle initiates multiple rounds of replication during each *S* phase, maintaining the episome at several dozen copies per cell. As a result, vectors based on the 2-micron origin are multicopy, whereas vectors based on chromosomal origins are low copy. All early yeast vectors were circular; more recently linear vectors resembling natural chromosomes have been constructed.

To achieve partitioning, a DNA molecule must possess a sequence that is recognized by the mitotic apparatus. Sequences responsible for partitioning are found in centromeres and ensure that each daughter cell receives one copy of each chromosome. If an episome does not contain a partitioning sequence, its inheritance becomes a simple matter of chance, and a certain fraction of daughter cells are formed without episomes. In the case of multicopy vectors the probability of loss by such segregation is rather low, and many such vectors were in fact constructed without any specific partitioning sequences.

To achieve selectability, a DNA molecule must possess a dominant marker, usually a gene that can be positively selected on a convenient medium. This need arises from the fact that the transfection process is inefficient, making it necessary to eliminate cells that have failed to assimilate vector molecules. In addition, cells lacking vector molecules that arise due to segregation during cell division can also be eliminated from cultures. An early example of a marker gene, still in wide use, is the

LEU2 gene of *S. cerevisiae*. The *LEU2* gene encodes one enzyme in the biosynthetic pathway for the amino acid leucine; cells with a *leu2*$^-$ genotype require a medium supplemented with leucine. Thus, a vector carrying the *LEU2* gene can be positively selected in *leu2*$^-$ cells using a medium lacking leucine. Another widely used marker is the URA3 gene, which encodes an enzyme in the pyrimidine biosynthetic pathway; this marker can be selected in *ura3*$^-$ cells on a medium lacking uracil. A general purpose yeast vector incorporating the above features is illustrated in Figure 5.4A.

The second breakthrough towards gene targeting was the realization that homologous recombination could occur between episomal and chromosomal sequences. One way to demonstrate this is simply to construct a vector lacking an origin of replication (see Fig. 5.4*B*). When such transfections were initially performed, colonies were nevertheless recovered on the selective plates. It was soon discovered that the transfected DNA molecules were being integrated into the chromosomal DNA,

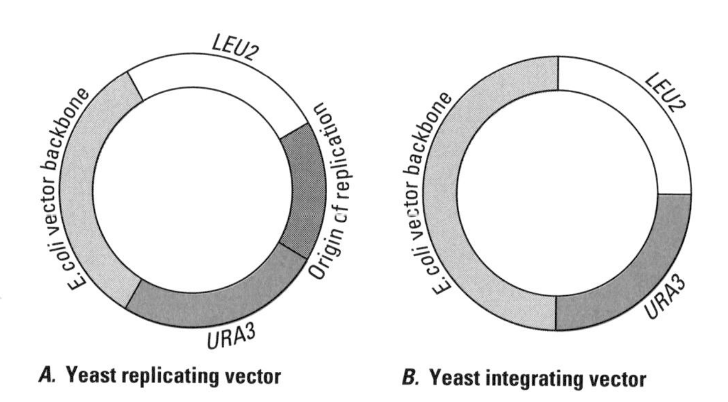

FIGURE 5.4 Two examples of yeast vectors. Most vectors are modular structures, composed of discrete segments of DNA, each of which confers a particular function. (*A*) Typical replicating vector, composed of four parts: a *LEU2* gene, a *URA3* gene, an origin of replication, and an *E. coli* vector backbone. The origin of replication can be obtained from either a chromosome or the 2-micron circle episome. The backbone is in fact a second vector containing an origin of replication and a selectable marker specific for *E. coli*. (Origins of replication and marker genes will usually function only in species closely related to the one from which they were derived.) The entire molecule can thus exist in both yeast and *E. coli* cells as an autonomously replicating extrachromosomal episome. The ability to propagate the vector in *E. coli* is necessary because amounts of DNA sufficient for in vitro manipulations such as the introduction of additional genes, sequence modifications, or sequencing can be obtained only from *E. coli* cells. (*B*) Typical integrating vector. It is identical to the vector in (*A*) except that the yeast origin of replication has been removed. This vector is thus incapable of autonomous replication in yeast cells; its replication in *E. coli*, however, remains unchanged.

in effect becoming a stable part of the genome and thus propagated from generation to generation. Genetic mapping experiments showed that integration always occurred at a chromosomal locus homologous to some part of the transfected DNA, in the case of the vector shown in Figure 5.4*B*, either at the *LEU2* or the *URA3* gene. This fact alone strongly implicated homologous recombination as the responsible process. Further detailed analyses confirmed that the integrated structures were consistent with a single homologous crossover (see Fig. 4.11). Integrations at non-homologous positions in the chromosomes can be recovered only extremely rarely.

The third breakthrough towards gene targeting was the demonstration that recombination with a chromosomal target sequence could be stimulated by simply introducing a double-stranded break in the homologous sequence on the vector. For example, in the case of the vector shown in Figure 5.4*B*, cutting the DNA in the *URA3* region prior to transfection would result in a majority of the recovered colonies containing an integration in the chromosomal *URA3* gene; conversely, cutting in the *LEU2* region would produce integrations predominantly in the *LEU2* gene. In essence, the integrations can be **targeted** to specific chromosomal sites. Observations such as these eventually led to the formulation of the double-strand-break repair model of homologous recombination, discussed in Chapter 4. In retrospect, it is easy to rationalize why cutting can target recombination: double-stranded ends can be converted to single-stranded tails that can in turn serve as efficient recombinogenic substrates via strand invasion of the homologous chromosomal regions.

Practical Aspects of Gene Targeting in *S. cerevisiae*

It is important to note that integration specificity is completely independent of the applied selection; in other words, integrations at both *URA3* and *LEU2* can be recovered with either a Ura$^+$ or Leu$^+$ selection being applied after transfection (see Fig. 5.5). This makes gene targeting a very powerful genetic tool that is completely general in its applications, since a dominant selectable marker can be used to target other genes. It is relatively easy to incorporate any gene—for example, a mystery gene *X* of unknown function—into an integrating vector. The beauty of gene targeting is that cutting within gene *X* prior to transfection, followed by selection for the positive marker in the vector, will result reproducibly in integration in the chromosomal gene *X* locus. The ability to integrate an

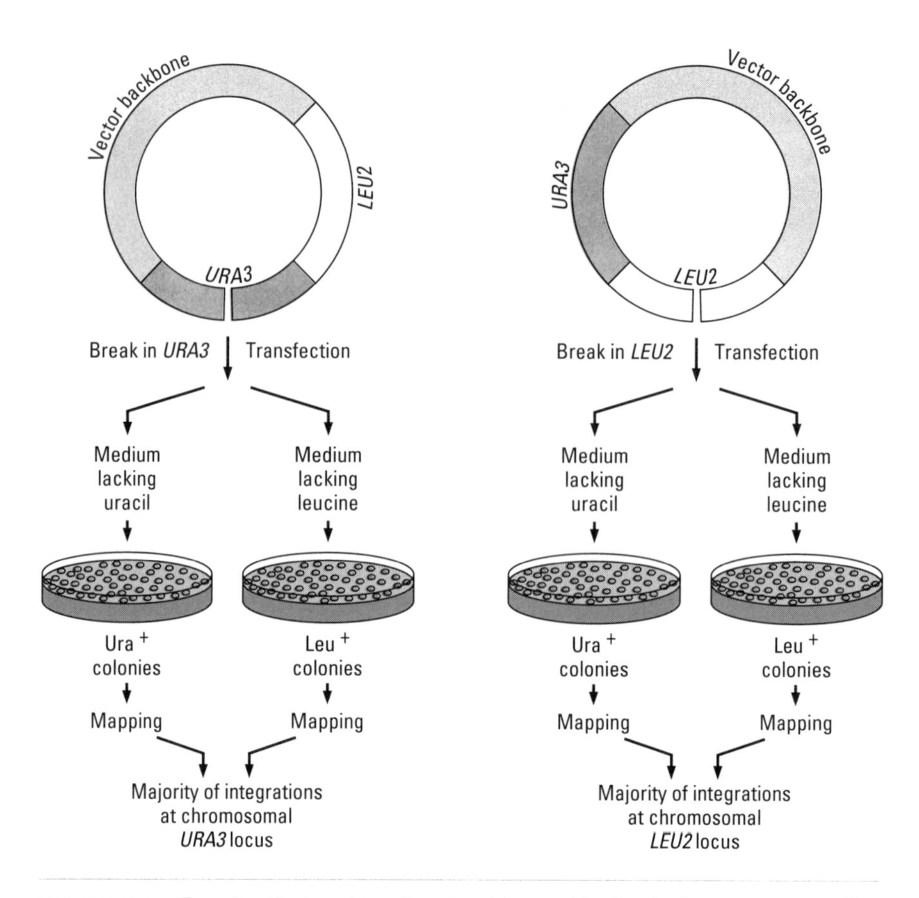

FIGURE 5.5 Targeting the insertion of vectors into specific sites in the yeast genome. The essence of gene targeting in yeast is that the site of integration is influenced primarily by the location of the site where the vector was linearized prior to transfection, not by the selection that is applied to obtain colonies.

exogenously introduced piece of DNA again and again into the homologous chromosomal locus provides the genetic engineer with the opportunity to introduce a variety of changes into the endogenous gene. The process consists of two steps: first, the desired mutation is constructed in the vector DNA in vitro, and second, the mutation is introduced into the chromosomal gene by gene targeting.

Four procedures for gene targeting are commonly used. The first two procedures were designed to introduce loss-of-function mutations into the chromosomal gene. In the genetic jargon these gene targeting methods are often referred to as gene disruptions or knockouts. In the first procedure, two independent mutations must be introduced into gene *X* in the vector in preparation for targeting (see Fig. 5.6*A*). The two mutations must be separated by a large enough portion of gene *X* to allow a homologous crossover between them. A relatively easy way to achieve this is to construct small deletions at the 5′ and 3′ ends of gene *X*, although other kinds of mutations, for example, frameshift or nonsense, can also be used. The key is that either the 5′ or the 3′ mutation alone must cause loss of function of gene *X*. The targeting is performed simply by cutting the vector in the middle of gene *X*, transfecting the DNA into yeast cells, and selecting for the marker in the vector (*LEU2*). The large majority of Leu^+ colonies will have integrated the vector into the chromosomal gene *X* by a single crossover (see Fig. 5.6*B*). The resultant structure contains two copies of gene *X* with the vector sequences sandwiched in between; the

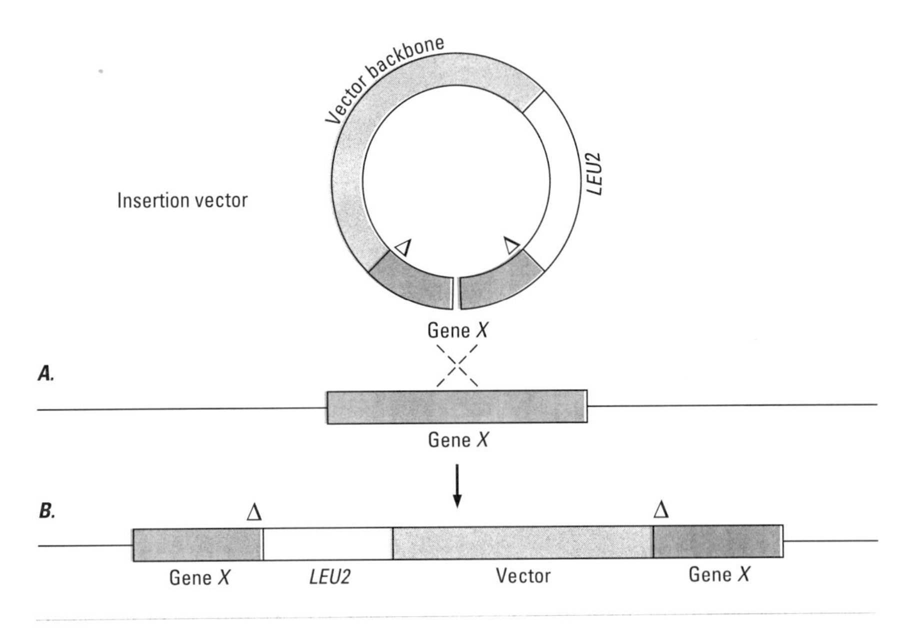

FIGURE 5.6 Gene disruption with an insertion targeting vector. (*A*) Starting construct containing a gene to be disrupted (gene *X*), a selectable marker (*LEU2*), and a vector backbone. Gene *X* in the vector has been modified on its 5′ and 3′ ends with two mutations, in this case small deletions, marked Δ in the drawing. Prior to transfection, a double-stranded break is introduced between the two mutations. Below the vector is shown a portion of the chromosome containing the recipient gene *X*. The position of the crossover is illustrated with dashed lines. (*B*) Product of the indicated single reciprocal crossover. Since no sequences are lost during the event, the integration results in a structure containing two copies of gene *X*, with the vector sequences, including the *LEU2* gene, sandwiched in between. One copy of gene *X* contains the 5′ deletion, while the other copy contains the 3′ deletion.

reason this causes a loss-of-function phenotype is that one of the copies contains the 5′ deletion while the other one has picked up the 3′ deletion.

This method was developed early and is not much used today. The reason is that two independent mutations must be constructed in gene *X* and tested to make sure that either one alone causes a loss-of-function phenotype. This is easy for the 5′ mutation but often not straightforward for the 3′ one, since slightly shortened proteins, or proteins with mutations at the very ends, often retain partial activity. This uncertainty can be seriously compounded by the fact that many knockout experiments are performed without prior knowledge of the loss-of-function phenotype. In fact, the very purpose of knockout targeting is often to discover what the loss-of-function phenotype is. Another less serious caveat about this procedure is that at a low frequency, a simple reversal of the crossover can regenerate a functional gene *X*.

In preparation for the second targeting procedure the vector is opened in the middle of gene *X* and the entire selectable marker gene is inserted, producing an insertion mutation. Often a portion of gene *X* is removed in addition to inserting the marker gene, producing a substitution mutation (see Fig. 5.7*A*). Since the coding region of gene *X* is thus grossly interrupted, it is unlikely that the encoded gene product will still be active. For this reason the step of testing for a loss-of-function phenotype prior to targeting is usually omitted. The targeting is performed by cutting the vector in two places, on either side of gene *X*, followed by the usual transfection and selection for Leu^+ colonies. The result is the substitution of the transfected DNA fragment into the chromosomal gene *X* by an event that can formally be represented as a double crossover on either side of the *LEU2* marker (see Fig. 5.7*B*). However, a gene conversion event initiated at one end of the transfected DNA and extending across the central region of nonhomology could also be invoked to account for the structure of the targeted locus.

This procedure, often referred to as one-step gene disruption, is currently the favorite with molecular geneticists working with *S. cerevisiae*. Although the mechanistic aspects of the recombination reaction may appear more complex, at least on paper, than the single reciprocal crossover of the previous method, targeted clones are recovered with approximately equal frequency using either procedure. The one-step gene disruption protocol, however, has several advantages. First, construction of the targeting vector is more straightforward. Second, the loss-of-function phenotype need not be explicitly tested prior to transfection. Third, the targeted gene *X* locus is stable and cannot revert to X^+.

The third and fourth procedures to be described here were designed to introduce more subtle mutations, including point mutations, into the chromosomal gene. In the genetic jargon these methods are sometimes

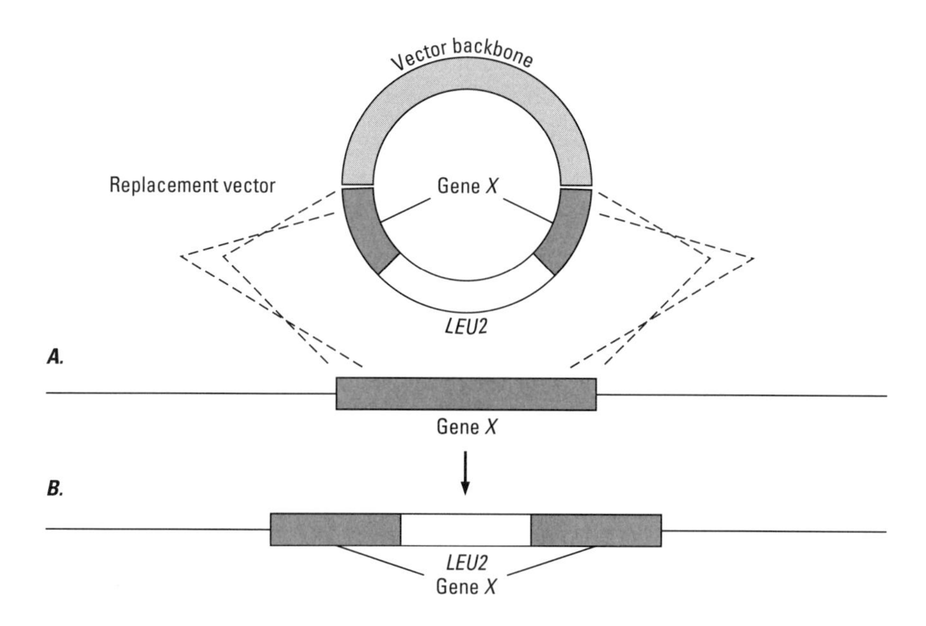

FIGURE 5.7 Gene disruption with a replacement targeting vector. (*A*) Starting construct containing a gene to be disrupted (gene *X*), a selectable marker (*LEU2*), and a vector backbone. Gene *X* in the vector has been interrupted in the middle by the insertion of the entire *LEU2* gene. Prior to transfection, two double-stranded breaks are introduced, one on either side of gene *X*. Below the vector is a portion of the chromosome containing the recipient gene *X*. The positions of two crossovers are illustrated with dashed lines. (*B*) Product of the indicated recombination events. In fact, although illustrated as two reciprocal crossovers, the event can also be modeled as a gene conversion initiated on either side of gene *X* and extending past the inserted *LEU2* gene. Two reciprocal crossovers would produce, in addition to the illustrated structure, a linear piece of DNA containing the chromosomal copy of gene *X*. This fragment, not shown in the drawing, would be subsequently either lost by cell division or degraded. Gene conversion, being a nonreciprocal event, would result directly in the degradation of the chromosomal copy of gene *X* and the substitution of corresponding plasmid sequences in its place. The vector backbone, not having any homology to the chromosomal locus, would not become involved in any recombination events. In either case, the end product is the replacement of the chromosomal locus with the disrupted, exogenously introduced copy of gene *X*.

referred to as gene replacements. The first is often called a pop-in/pop-out replacement. The targeting is based on a single crossover to produce an integrated structure, followed by a reversal of the crossover. In preparation for targeting, the desired point mutation is introduced into gene *X* in the vector (see Fig. 5.8*A*). The mutation is placed well inside the region of homology with the chromosome so that a single crossover on either the 5′

FIGURE 5.8 (*Right*) Gene replacement with a pop-in/pop-out targeting vector. (*A*) Starting construct containing a gene to be disrupted (gene *X*), a selectable marker (*LEU2*), and a vector backbone. Gene *X* in the vector has been modified by the introduction of a single point mutation, marked * in the drawing. Prior to transfection, a double-stranded break is introduced between the point mutation and the end of gene *X*. Below the vector is shown a portion of the chromosome containing the recipient gene *X*. The position of the crossover is illustrated with dashed lines. (*B*) Product of the indicated single reciprocal crossover. As in Figure 5.6, the integration results in a structure containing two copies of gene *X*, with the vector sequences, including the *LEU2* gene, sandwiched between. One copy of gene *X* contains the point mutation, while the other copy remains unchanged. (*C*) One way in which recombination between the two copies of gene *X* can resolve the structure. If the two copies of gene *X* become aligned by looping out the intervening vector sequences, a single reciprocal recombination event can restore the original configuration, namely a linear chromosome and a circular episome each containing one copy of gene *X*. Whether the point mutation in gene *X* ends up in the chromosome or the episome is determined by the precise location of the crossover; both alternatives are illustrated in the bottom part of *C*. Since the recombination involves two regions of a single chromosome, it is often referred to as intrachromosomal recombination. (Although not illustrated in this figure, another way to resolve the structure containing the two copies of gene *X* is by unequal crossing over shortly after this region of the chromosome has been replicated. This process invokes a misalignment of, for example, the rightward copy of gene *X* on one of the replicated chromosomes with the leftward copy of gene *X* on the other replicated chromosome, followed by a single reciprocal crossover. Following mitosis, one daughter cell would inherit one copy of gene *X* and the other would inherit three copies.)

or the 3′ side is possible. The targeting is performed by cutting the vector in gene *X* on one side of the point mutation, transfecting the DNA into yeast cells and selecting for the marker in the vector (*LEU2*). As expected, the vector integrates into the chromosome by a single crossover (see Fig. 5.8*B*).

As in the first procedure (see Fig. 5.6), the integrated structure contains a duplication of gene *X* with the vector sequences sandwiched between. The point mutation is present in one of the copies of gene *X* while the other copy remains unchanged; as a result, the phenotype of the targeted clones is still X^+. The second step involves the reversal of the integration event by a single crossover between the two duplicated copies of gene *X* (see Fig. 5.8*C*). This process is sometimes referred to as looping out. Since the point mutation is internal to the region of homology, the crossover to generate the loop-out can occur on either side of it. If the looping-out crossover occurs on the same side as the integrating crossover, the chromosomal copy of gene *X* will remain unchanged, and the vector will be regenerated (see Fig. 5.8*C*-I). On the other hand, if the looping-out crossover occurs on the opposite side of the point mutation from the

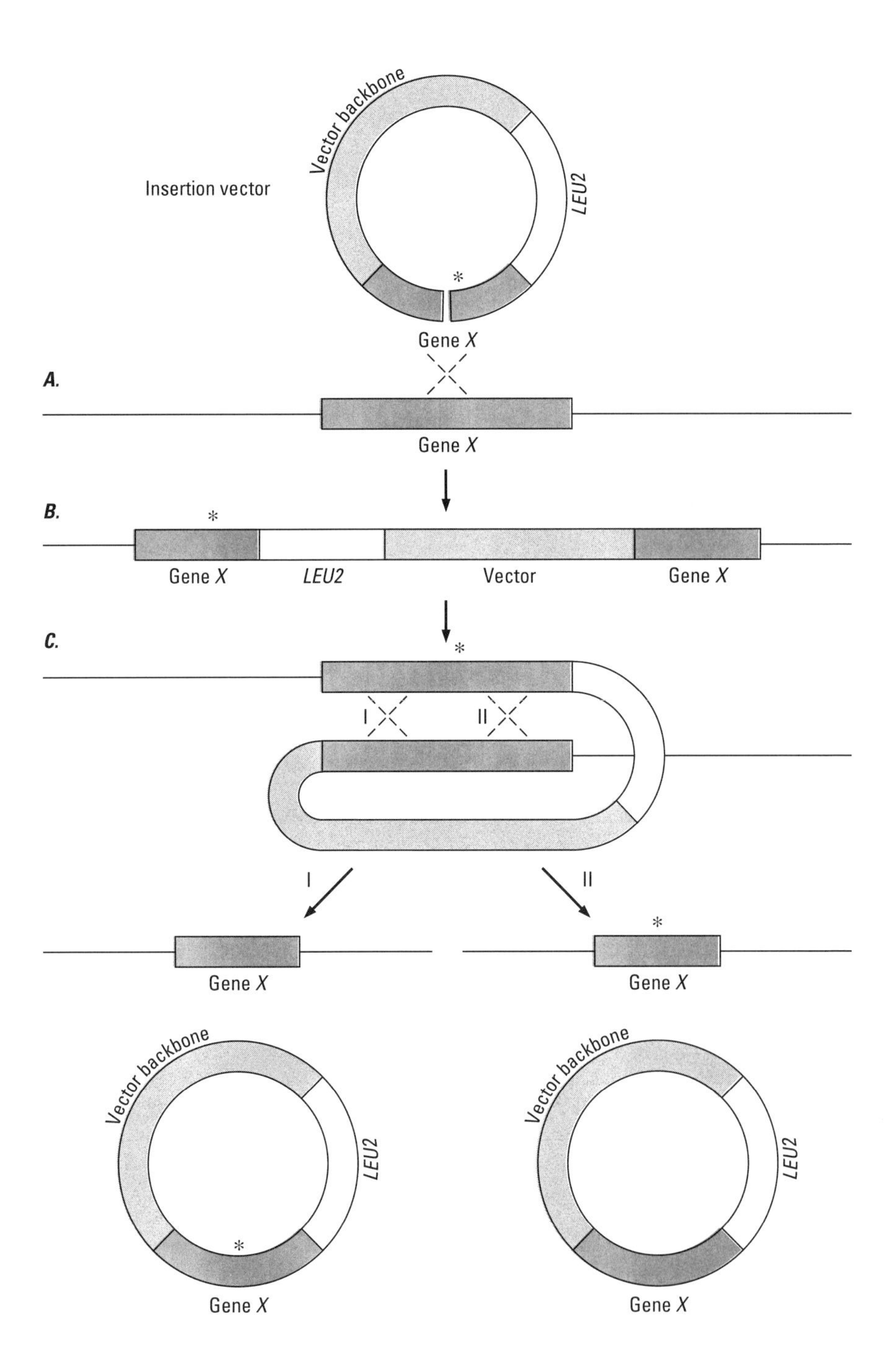
Insertion vector
Vector backbone
LEU2
*
Gene X
A.
Gene X
B.
*
Gene X
LEU2
Vector
Gene X
C.
*
I
II
I
II
Gene X
*
Gene X
Vector backbone
LEU2
*
Gene X
Vector backbone
LEU2
Gene X

integrating crossover, the mutation will remain in the chromosome, and the vector will pick up the normal copy of gene *X* (see Fig. 5.8*C*-II).

This procedure is more sophisticated than the first two, and as such it has its special advantages and disadvantages. The most important point is its versatility: it allows the isolation of targeted clones containing a virtually unlimited spectrum of mutations. Second, other than the introduction of the desired mutation, the chromosomal locus remains unchanged, since neither vector sequences nor selectable marker genes are left behind. Third, the introduced mutation is stable. The disadvantage is the added complexity that results from the need for the second crossover. Cells that have looped out the vector will give rise to Leu^- clones, because the vector lacks its own origin of replication. To recover the second crossover, one approach is to simply grow the cells under nonselective conditions (on medium containing leucine), followed by a screening procedure for Leu^- colonies. Since the loop-out is a relatively rare event, this method is time consuming. Another approach is to include a negative marker in the vector. In other words, a positive selection is applied during the "in" phase of targeting, followed by a negative selection during the "out" phase to actively select for cells that have looped out and segregated the vector. This method is much more convenient but necessitates the inclusion of a negative marker in the vector. The *LEU2* gene cannot be used as a negative selectable marker, but convenient negative selections have been developed for some other genes, for example, *URA3*. The starting vector must thus be modified to include such a marker. The final step in the pop-in/pop-out protocol is a screen for the incorporation of the desired mutation into the chromosomal copy of gene *X*. Since the crossover during the "out" step can occur on either side of the mutation, clones will be recovered that contain the original or the modified version of gene *X*. If the change being introduced into the chromosome is very small, for example, a single base pair change, the final screening can be laborious.

The second gene replacement procedure is often referred to as a direct replacement. The method in essence consists of two one-step gene disruption targeting experiments performed in succession, the first one employing a positive selection and the second one a negative selection. The preparation for targeting is the same as for a one-step gene disruption (see Fig. 5.7), except that the marker used to interrupt gene *X* should allow both a positive and a negative selection. One such marker is *URA3*, whose presence can be selected in *ura3*$^-$ cells on a medium lacking uracil, and whose absence can be selected on a medium containing 5-fluoroorotic acid, a toxic analog (see Fig. 5.9A). The targeting is performed by cutting the vector on both sides of gene *X* followed by selection for Ura^+ colonies. The result is the substitution of the transfected DNA fragment into the chromosome and the disruption of gene *X* (see Fig. 5.9*B*). The second step

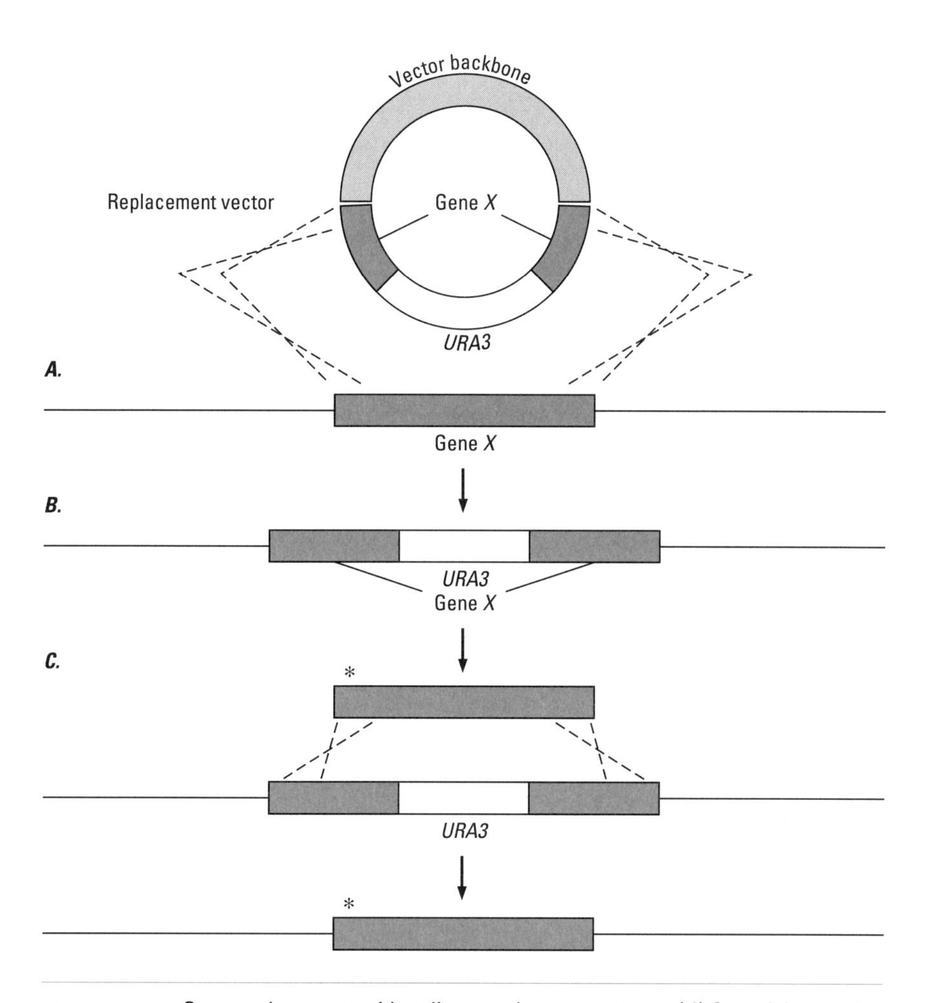

FIGURE 5.9 Gene replacement with a direct replacement vector. (*A*) One of the starting constructs, containing a gene to be disrupted (gene *X*), a selectable marker (*URA3*), and a vector backbone. The first part of this procedure is analogous to gene disruption with a replacement targeting vector (see Figure 5.7). Gene *X* in the vector is first interrupted with the *URA3* gene. Prior to transfection, two double-stranded breaks are introduced, one on either side of gene *X*. Below the vector is shown a portion of the chromosome containing the recipient gene *X*. The positions of the two crossovers are illustrated with dashed lines. (*B*) Product of the indicated recombination events. (*C*) Second part of the procedure. A second construct, containing gene *X* in a vector backbone, is first modified by the introduction of a single point mutation, marked * in the drawing. Prior to transfection, two double-stranded breaks are introduced, one on either side of gene *X* (the liberated vector backbone is not included in the drawing). After transfection, the cells are selected for a Ura^- phenotype. The positions of the two crossovers are illustrated with dashed lines. The product of the indicated recombination events is shown at the bottom of the figure.

involves transfection with gene X containing the desired point mutation, followed by selection for Ura$^-$ colonies. Ura$^-$ colonies can arise either by the occurrence of spontaneous *URA3*$^-$ mutations in the inserted URA3 gene or by replacement of the *ura3* gene by substitution of the transfected fragment of gene *X* (see Fig. 5.9*C*). The frequency of spontaneous *ura3*$^-$ mutations is in the range of 1 per million (10^6) cells, which is high enough to occasionally present a problem. This is because the overall frequency of the sought gene targeting events, which is actually the frequency of transfection multiplied by the frequency of homologous recombination, can be low enough to be in the same range as the frequency of *ura3*$^-$ mutations. The method that has been used to overcome this problem is to cotransfect the linear targeting module with a circular replicating vector containing a distinct positive marker (for example, the vector shown in Figure 5.4*A*). The method is based on the observation that during the transfection process, only a small fraction of the cells become competent to take up DNA, but those that do take up many DNA molecules. Thus, selecting for the positive marker present on the cotransfected vector eliminates most of the nontransfected cells, making the overall frequency of the sought gene targeting events closer to the frequency of homologous recombination. The two transfected DNAs should preferably contain minimal homology so that they will not recombine with each other.

The direct gene replacement targeting strategy has two important advantages. First, it is more efficient at allowing numerous replacements to be performed at a single locus. The pop-in/pop-out strategy has to be performed in its entirety for each point mutation to be introduced into the chromosome. With the direct gene replacement strategy, the step of introducing the negative marker into the chromosomal locus need be performed only once. Second, unlike the pop-in/pop-out strategy, every homologous recombination event in the second step should result in the incorporation of the point mutation into the chromosome. The only significant source of background is the spontaneous mutations in the negative marker.

The Special Fascination of Gene Targeting

The development of gene targeting led to a virtual revolution in the molecular genetics of *S. cerevisiae*. The major application has been the construction of loss-of-function mutants. Why is this so?

In the effort to discover what makes cells tick, molecular biologists pose one question over and over again: What does this gene product do?

Biochemical analysis is one way to shed light on the issue; after the appropriate battery of tests is applied, it is usually possible to define, in some fashion, an **activity.** One protein may be shown to catalyze, for example, a particular step in a biosynthetic pathway, another may turn out to be a nuclease and thus implicated in some aspect of DNA metabolism, yet another may turn out to be a structural component of some tissue, and so forth. This process has two shortcomings. First, the analysis is usually performed in the test tube; thus, cells must be broken open before the experiment can start. Historically, this process has proven to be prone to artifacts, for the simple reason that breaking cells apart can introduce changes in protein activities. This is understandable, since the environment in a test tube can never exactly duplicate the one inside a living cell. Second, even if an activity can be faithfully reproduced in vitro, questions remain: What does this activity really do for the cell? Why is it important? What would happen if it were not there?

Fortunately, a genetic analysis can provide the answer to the last question, and through a study of the loss-of-function phenotype, some answers may be found even for the others. For an enzyme involved in the biosynthesis of, for example, leucine, the answer may be relatively straightforward: The cell can no longer make leucine. For an exonuclease, the answer may be more revealing and implicate the activity in DNA replication or recombination, or both, or even neither.

In short, loss-of-function phenotypes force living cells to indicate why they need a particular gene. It is, of course, often not easy to decipher what cells are trying to say, but that is a different matter. Classical microbial genetics, as discussed earlier, mandates that phenotype goes before genotype. In other words, the experimenter formulates a selective medium, the medium selects for a particular phenotype, and any genotype that offers a selective advantage will score as a colony. Thus, the experimenter is often left with the task of sorting for the genotypes he or she set out to isolate. Of course, finding a genotype that correlates with an interpretable phenotype as well as a change in a biochemical activity can constitute a well-argued case.

The special fascination of gene targeting is that it throws this age-old paradigm on its head. An experimenter can start with a gene as a piece of DNA in the test tube and without knowing anything else about it, construct a guaranteed loss-of-function genotype. All that remains is to ask, What is the consequent phenotype? For this reason, the advent of gene targeting is sometimes said to have ushered in the age of reverse genetics. It is, of course, somewhat absurd to speak of "forward" and "reverse" genetics, since genetics is, after all, just genetics; the catchword is simply meant to emphasize the profound change in experimental possibilities that gene targeting has brought.

Nonhomologous Recombination Is the Chief Obstacle to Gene Targeting in Mammalian Cells

Why has gene targeting been so difficult in mammalian cells? The primary reason, in a nutshell, is the occurrence of **nonhomologous recombination.** The process of nonhomologous recombination, sometimes also referred to as **illegitimate recombination,** is conceptually much simpler than homologous recombination: A transfected piece of DNA becomes ligated, end to end, into a double-stranded break introduced somewhere in the genome. The reaction appears to be quite random and proceeds without any apparent consideration for homology between the two DNA molecules.

Consider what happens then, for example, if a one-step gene disruption targeting experiment is attempted in mammalian tissue culture cells. Construction of the vector is easy, since all that is needed is the target gene from the correct species and an appropriate positive selectable marker. Neither the transfection nor the recovery of sufficient numbers of positively selected colonies is a problem. The disappointment comes when the fate of the transfected DNA is more closely examined: All colonies contain random integrations at nonhomologous sites throughout the genome. This, of course, is because the selective medium selects only for the phenotype; if the positive marker gene is adequately expressed and integrated into a nonhomologous site, which it usually is, that cell will score as a colony.

What actually happens during transfection of DNA into a mammalian cell? Several transfection methods are currently available; in essence they are all contrivances of one sort or another to get the DNA into the cell across the cellular membrane, which under normal circumstances is an effective barrier. Once the DNA is inside the cell, by an unknown process it eventually ends up in the nucleus (see Fig. 5.10). Most of the DNA never makes it and is degraded along the way. A considerable fraction of the DNA that appears in the nucleus has suffered some degree of endonucleolytic or exonucleolytic damage. Once inside the nucleus, any genes present in the DNA begin to be transcribed. The whole process is relatively rapid; within 12 hours of transfection the gene products encoded by the transfected DNA begin to appear. The transfected DNA is also assembled with nucleosomes and undergoes several rounds of replication during each *S* phase if it contains an origin.

In the absence of an origin of replication, the DNA is rapidly lost by dilution due to cell division as well as ongoing nucleolytic degradation. The expression of gene products from vectors lacking origins thus decays within a few days; this burst of activity is often referred to as transient expression. The fraction of the total cells subjected to transfection that

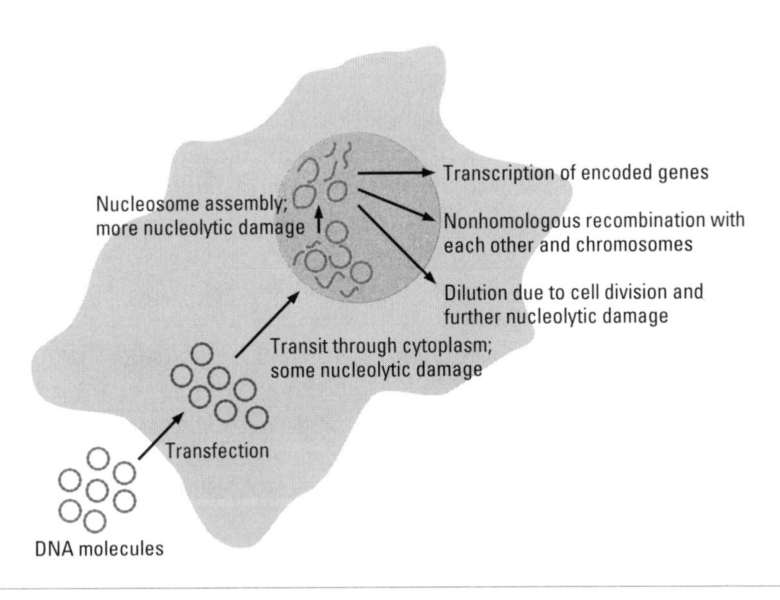

FIGURE 5.10 Intracellular fates of transfected DNA.

display transient expression, which is an indicator of the efficiency with which the DNA penetrates into the cell and is transported to the nucleus, is highly variable. This parameter depends on both the type of cell being used and the method of transfection; for unknown reasons, some cell types can be highly refractory to transfection. A frequency of transient transfection in range of a few percent is a good average, but with several cell types efficiencies as high as 20 to 30 percent have been achieved.

Nonhomologous recombination of the transfected DNA also occurs within this window of transient expression. In fact, the transfected DNA molecules interact rather avidly with each other, often resulting in a complex mixture of rearrangements. Nonhomologous recombination with chromosomal DNA can be observed within 24 hours of transfection, but it occurs in only 1 to 10 percent of the cells that successfully transport DNA into their nuclei. Integration into the genome by nonhomologous recombination can lead to clones of cells that stably express gene products encoded on the transfected DNA.

Although nonhomologous recombination may appear primitive, essentially nothing is known about the underlying molecular mechanisms. Interestingly, it does not occur in *E. coli* and *S. cerevisiae.* Although chromosomal rearrangements such as deletions, inversions, and duplications in these organisms are, by definition, illegitimate recombination events, a substantial fraction of them may in fact happen by homologous recombination through short regions of homology—on the order of six to ten base pairs—that occur simply by chance in otherwise nonhomologous regions of the genome. In addition, illegitimate events in *E. coli* and *S. cerevisiae* are very rare and thus difficult to study.

The position on the evolutionary tree appears to roughly correlate with the propensity for nonhomologous recombination. Some higher fungi and protozoans, for example, display a nearly equal mix of homologous and nonhomologous recombination. Invertebrates, such as the nematode worm *Caenorhabditis elegans* or the fruit fly *Drosophila melanogaster,* already possess high levels of nonhomologous recombination. In mammalian cells, where the issue of nonhomologous versus homologous recombination has been examined most extensively, the ratio is, on the average, between 10,000 : 1 and 1,000 : 1. This ratio reflects only stable events; neither nonhomologous nor homologous events that occur between the transfected molecules during the transient phase but fail to be integrated into the genome are included in the figure.

Why should higher organisms possess nonhomologous recombination activity? It must be emphasized that the high levels of nonhomologous recombination discussed involve exogenous DNA newly introduced into cells. Chromosomal DNA sequences during a normal, unperturbed life cycle appear in contrast quite stable; this makes sense since otherwise genomes would be undergoing constant scrambling. The existence of nonhomologous recombination is thus somewhat of a mystery; one explanation that may be plausible is that it is a last fallback mechanism to preserve the physical integrity of chromosomes. In a chromosome that suffers a double-stranded break, all the DNA distal to the centromere will be lost at the subsequent cell division. If everything else fails, it would appear more advantageous for a cell to hang on to the acentric piece of DNA by recombining it nonhomologously somewhere else in the genome than to lose it altogether. It has thus been sometimes said that mammalian cells "hate" free DNA ends and try to ligate them as fast as possible. It is possible that cells may mistake exogenous DNA for broken-off pieces of their own chromosomes and try to hang on to it by nonhomologous recombination.

Gene Targeting Is a Feasible Goal in Mammalian Cells

The occurrence of homologous recombination between viral genomes replicating in mammalian cells was already established in the 1960s. These observations implied the existence of cellular recombination enzymes, since some viruses are too small to encode their own. The picture remained somewhat murky until homologous recombination between nonviral DNAs could be quantitatively demonstrated, which became possible in the late 1970s and early 1980s as transfection methods into mammalian cells were improved and, most important, good positive selectable markers became available.

As mentioned earlier, the lack of selectable markers that can be effectively used in a wide variety of cultured mammalian cells has been and continues to be a serious impediment to genetic manipulations. Even today, only a handful of marker genes are available. The most widely used markers are in fact bacterial genes that have been adapted for expression in mammalian cells. The current favorite is a gene designated *neo,* which encodes an enzyme that confers resistance to the neomycin class of antibiotic drugs. Neomycin inhibits translation in bacterial cells, and a related drug called G418 inhibits translation in mammalian cells. Although the bacterial enzyme has evolved to be active against neomycin, it will also recognize and detoxify G418. Another bacterial gene that has been adapted for use as a marker in mammalian cells confers resistance to the drug hygromycin.

A classic early experiment was to construct two vectors, one containing a mutation in the 5′ end of the *neo* gene, and the other a mutation in the 3′ end. Each mutation caused a loss-of-function phenotype; thus neither vector alone could confer resistance to G418. A single crossover between the two vectors, however, through the homologous central regions of the two *neo* genes, would result in one normal and one doubly mutant gene (see Fig. 5.11). The power of this approach is the absence of any background. First, if the two *neo* mutations are small deletions, neither can revert to Neo$^+$; second, mammalian cells have never been observed to spontaneously acquire G418 resistance. Thus, all G418-resistant colonies recovered following a transfection of the two vectors are due to homologous recombination events.

Initial experiments using this system were to simply cotransfect the two vectors, thus allowing them to recombine during the transient phase in an extrachromosomal state. The frequency with which stable G418 colonies are recovered in such an experiment is actually a product of several values: the frequency of transient transfection, the frequency of homologous recombination, and the frequency of nonhomologous recombination of the reconstructed *neo* genes with the genome. Another way to perform the experiment is to allow the vectors to recombine during the transient phase but subsequently recover them from the nuclei and transfect them back into *E. coli* to score for the reconstruction of the *neo* gene. The contribution of the frequency of nonhomologous recombination in the previous experiment can thus be eliminated. One problem

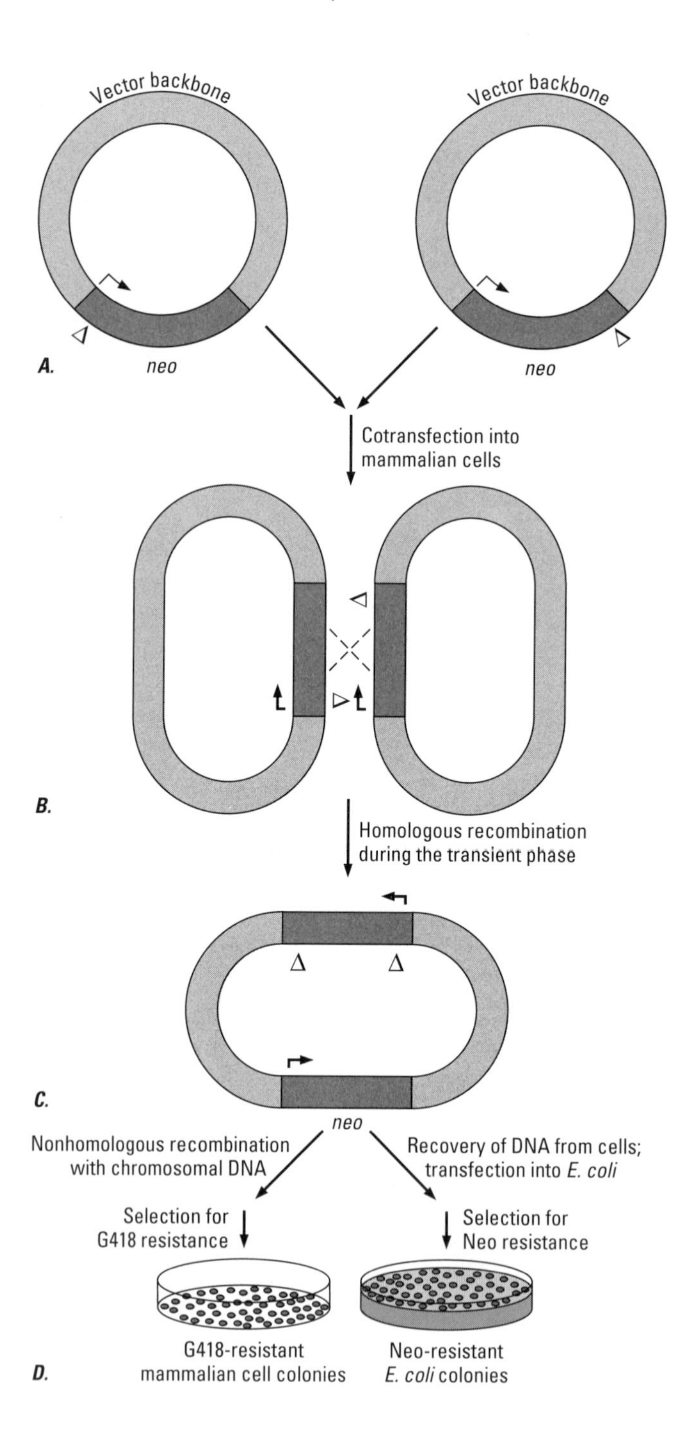
Vector backbone
Vector backbone
A.
neo
neo
Cotransfection into
mammalian cells
B.
Homologous recombination
during the transient phase
Δ
Δ
C.
neo
Nonhomologous recombination
with chromosomal DNA
Recovery of DNA from cells;
transfection into E. coli
Selection for
G418 resistance
Selection for
Neo resistance
G418-resistant
mammalian cell colonies
Neo-resistant
E. coli colonies
D.

FIGURE 5.11 *(Left)* Extrachromosomal homologous recombination between two vectors transfected into mammalian cells. (*A*) Two starting vectors, one containing a *neo* gene with a 5′ deletion, the other a *neo* gene with a 3′ deletion. The deletions are indicated as Δ in the figure; each mutation alone is sufficient to inactivate the Neo gene product. The direction of transcription of the *neo* genes in the vectors is indicated with an arrow. (*B*) The two vectors positioned with the *neo* genes aligned for homologous recombination; a single reciprocal crossover is indicated with dashed lines. (*C*) Product of the recombination: a circle containing two copies of the *neo* gene as well as two vector backbones. One of the *neo* genes is free of mutations and thus encodes a functional gene product, while the other gene has picked up both of the deletions. (*D*) Two alternative procedures for scoring the frequency of the recombination reaction by recovering the correct products as colonies of cells. The first assay involves simply adding the drug G418 to the medium. This requires the products of the homologous recombination reaction, shown in *C*, to subsequently recombine nonhomologously with the genome in order to be stably inherited. The second assay involves recovering the DNA from the nuclei of the mammalian cells and transfecting it back into *E. coli* cells, which are subsequently selected for resistance to neomycin in order to recover the products of the correct homologous recombinations.

with this approach is that it must be carefully shown that the homologous recombination is actually taking place inside the mammalian cells and not in the *E. coli* cells during the last phase of the experiment. Yet another way to perform the experiment, if the frequencies of homologous recombination are sufficiently high, is to detect the products by direct physical assays without resorting to a genetic selection.

To everyone's delight, it was immediately apparent that homologous recombination was taking place at surprisingly high frequencies! The second important observation was that linearization of one of the vectors in the region of *neo* homology prior to transfection enhanced the frequency of recombination. Although the levels of stimulation reported by various investigators were often variable, an approximately tenfold enhancement is a good average. This was comforting because it was in line with observations from gene targeting work in *S. cerevisiae*. In some cases it was shown that up to 20 percent of the DNA molecules that successfully made it into the nuclei of mammalian cells became involved in homologous recombination events!

To model gene targeting more directly, subsequent work turned to the analysis of recombination between transfected DNA and homologous chromosomal sequences. To make such experiments possible, one of the mutant *neo* vectors was first stably integrated into the genome by nonhomologous recombination, in essence creating an artificial chromosomal locus whose targeting could be directly selected (see Fig. 5.12). A clonal population of such cells was subsequently transfected with the second

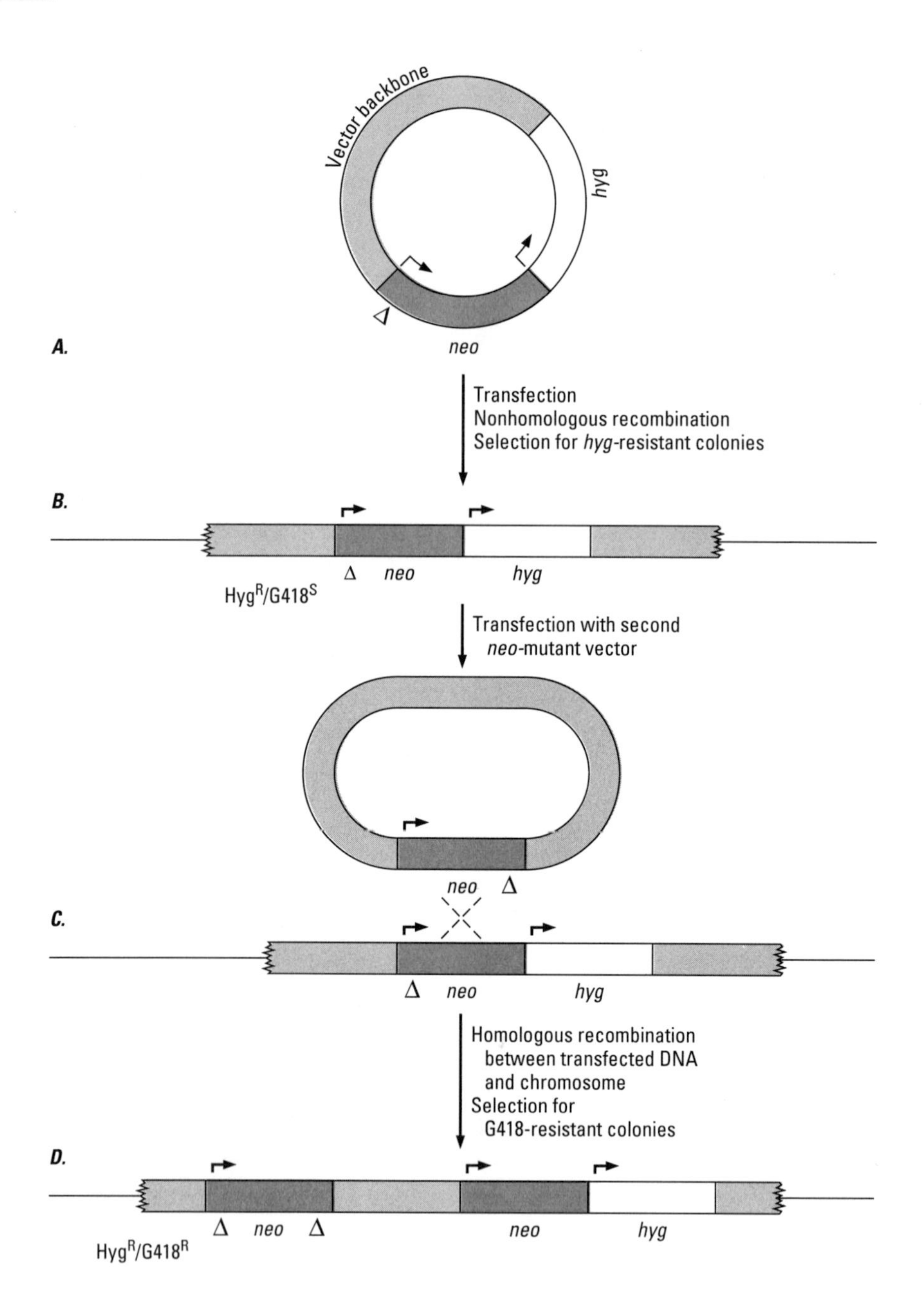
Vector backbone
hyg
neo
A.
Transfection
Nonhomologous recombination
Selection for hyg-resistant colonies
B.
Δ neo
hyg
HygR/G418S
Transfection with second
neo-mutant vector
neo Δ
C.
Δ neo
hyg
Homologous recombination
between transfected DNA
and chromosome
Selection for
G418-resistant colonies
D.
Δ neo Δ
neo
hyg
HygR/G418R

FIGURE 5.12 (*Left*) Homologous recombination between an extrachromosomal vector and a vector stably integrated into the genome. (*A*) One of the two starting vectors, containing a *neo* gene with a 5′ deletion, an *hyg* gene (conferring resistance to the drug hygromycin), and a vector backbone. The deletion is indicated as Δ in the figure and is sufficient to inactivate the Neo gene product. The *hyg* gene encodes a functional gene product. The direction of transcription of the two genes is indicated with arrows. (*B*) Product of a nonhomologous recombination reaction in which the vector was opened in the backbone region and integrated into a random chromosomal site. The phenotype of a cell containing this structure would be hygromycin-resistant and G418-sensitive. (*C*) Introduction of the second vector, containing a 3′ deletion in the *neo* gene. The incoming vector is shown positioned next to the chromosome and aligned for homologous recombination; a single reciprocal crossover is indicated with dashed lines. (*D*) Product of the homologous recombination. As in Figure 5.11, one of the *neo* genes is free of mutations and thus encodes a functional gene product, while the other gene has picked up both of the deletions. The phenotype of a cell containing this structure would be G418-resistant as well as hygromycin-resistant.

mutant *neo* vector, followed by selection for G418-resistant colonies. The frequency of resistant colonies was low but fortunately not zero. Moreover, cutting the vector in the region of *neo* homology prior to transfection increased the recovery of resistant colonies, and a detailed physical analysis of the integrated structures showed that they were indeed produced by homologous recombination. When such experiments were repeated with a number of different clones of cells, each containing the mutant *neo* gene at a distinct chromosomal location, each target locus recombined with a somewhat different frequency. This was not altogether unexpected, since it was known from classical genetic studies that not all regions of the genome are equally recombinogenic. One way to explain such **position effects** is to invoke the existence of recombinational hot spots, perhaps analogous to *E. coli* Chi sites, where recombination can be initiated. Mutant *neo* genes that fortuitously integrated close to such sites would then be subject to gene targeting with a relatively high efficiency, while those that were far away might be quite inaccessible.

Recombination experiments with the *neo* genes established the foundations for the ultimate success of gene targeting in mammalian cells. First, these experiments demonstrated unequivocally the existence of homologous recombination between transfected DNA molecules and chromosomal sequences, showing that gene targeting was a feasible goal. Second, the frequencies of gene targeting, as well as the magnitude of the background due to nonhomologous recombination were quantitated. Third, the relative ease with which the direct selections could be performed allowed some important parameters to be established, for example, the stimulation of recombination by double-stranded breaks in the region of homology.

The next model system made use of a chromosomal gene whose loss of function could be directly selected. In this way homologous recombination events could still be directly selected, but the target was a real cellular gene as opposed to an artificial target introduced into unknown locations by nonhomologous recombination. The gene used encodes the enzyme hypoxanthine phosphoribosyl transferase (Hprt) and is involved in the utilization of purines from the medium. Cells can synthesize nucleotides from very simple precursors, but if purines are added to the medium they will be preferentially used. This makes sense, since it saves the energy of having to synthesize them from scratch. The loss of Hprt activity can be directly selected using a toxic analog of guanine, 6-thioguanine. An $Hprt^+$ cell will incorporate 6-thioguanine into its RNA and DNA and die; an $Hprt^-$ cell will simply switch to the synthesis of its own purines and survive. Since the *hprt* gene is found on the X chromosome, it is a haploid marker in male cells. In other words, a single mutation, or gene targeting event, can cause an $Hprt^-$ phenotype.

The use of the *hprt* gene as a model brought several important advances. First, it was shown that natural mammalian genes could be targeted with roughly the same efficiency as artificial *neo* targets. Second, the two basic methods of gene disruptions—insertion (single crossover constructs) and replacement (double crossover constructs) were compared and found to recombine with equivalent frequencies (see Figs. 5.13 and 5.14). Third, additional parameters were established, of which the most important was that an increase in the length of homology between the transfected DNA and the chromosome can dramatically increase the frequency of recombination. In previous studies the length of homology was limited to the distance between the two *neo* mutations (see Fig. 5.11); using the *hprt* gene, homologies of up to 9000 base pairs were tried. Fourth, the possible existence of hot spots was again encountered. Targeting frequencies at one end of the *hprt* gene that were found to be significantly higher than those found at the other end of the gene.

Quantitative Determination of Targeting Frequencies Is Difficult

The calculation of targeting frequencies and, particularly, the ratios with respect to the ever-present background of nonhomologous recombination are subject to several sources of uncertainty. This fact makes the data from different laboratories sometimes difficult to compare. Part of the problem is to decide on the units in which frequencies are to be expressed. One possibility is to express the genetic events, either homologous or nonhomologous, as the

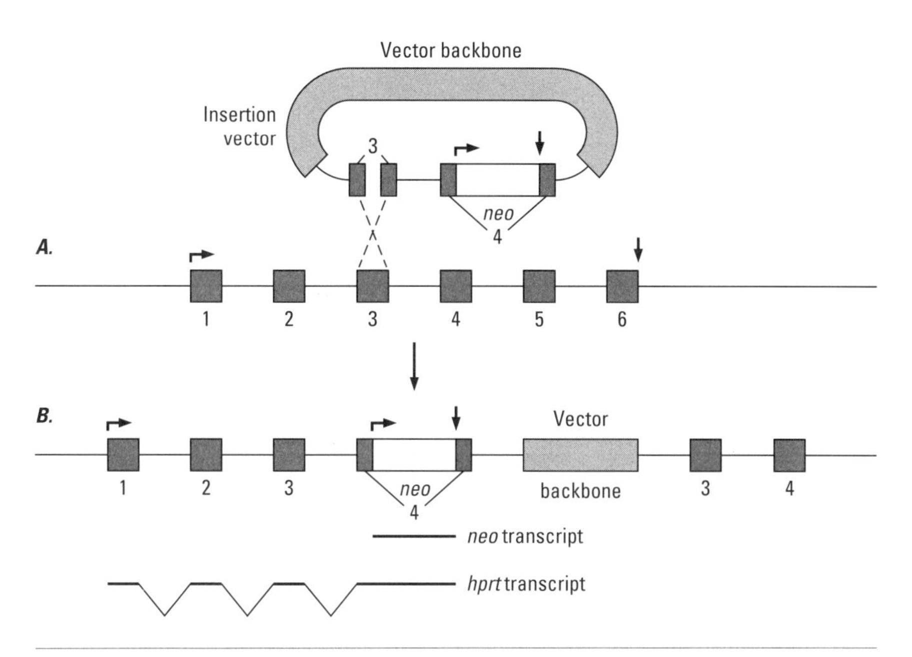

FIGURE 5.13 Disruption of the *hprt* gene in mammalian cells using an insertion vector. (*A*) Incoming targeting vector positioned above the chromosomal *hprt* gene and aligned for homologous recombination. The *hprt* gene, like most genes in mammalian cells, is composed of introns and exons. Exons are drawn as boxes and introns are shown as solid lines. Only 6 of the 9 *hprt* exons are shown. The initiation of transcription at the start of exon 1 is indicated with an arrow. The targeting vector has the *neo* gene inserted in the middle of exon 4. The *neo* gene contains its own promoter, and the direction of transcription is indicated with an arrow. Prior to transfection, a double-stranded break is introduced in the targeting vector within *hprt* sequences; the cut is shown in exon 3. A single reciprocal crossover is indicated with dashed lines. (*B*) Product of the homologous recombination. As in Figure 5.12, the entire targeting vector has been incorporated into the chromosome, resulting in the duplication of some *hprt* sequences. The components of the recombinant structure are, in linear order, normal chromosomal *hprt* sequences up to exon 3, intron 3, exon 4 interrupted with the *neo* gene, portion of exon 4, vector backbone, portion of intron 2, exon 3, followed by the rest of the *hprt* gene. The transcripts expected from this structure are indicated at the bottom. One transcript should be initiated at the *neo* promoter and encode a functional gene product; thus, the targeted cell would display a G418-resistant phenotype. A second transcript should be initiated at the *hprt* promoter; following splicing to remove the introns, the mRNA would retain the insertion of *neo* sequences in exon 4. Thus, the encoded Hprt gene product would be nonfunctional, and the cell would display an $Hprt^-$ phenotype.

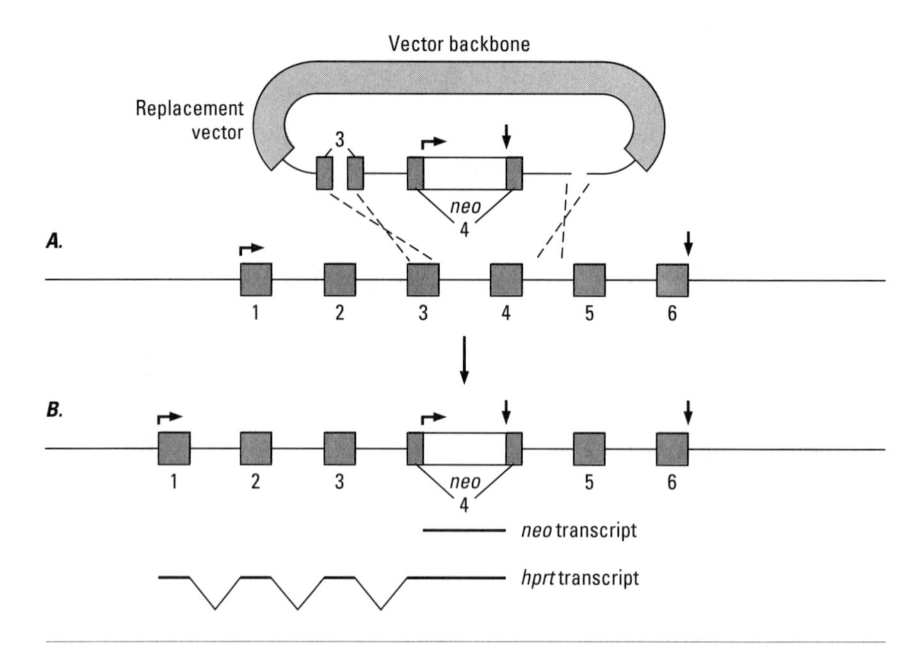

FIGURE 5.14 Disruption of the *hprt* gene in mammalian cells using a replacement vector. (*A*) Incoming targeting vector, identical to the insertion vector in Figure 5.13, positioned above the chromosomal *hprt* gene and aligned for homologous recombination. Refer to Figure 5.13 for details of *hprt* gene and targeting vector structure. Prior to transfection, two double-stranded breaks are introduced into the targeting vector within *hprt* sequences, one on either side of the *neo* insertion. Two reciprocal crossovers are indicated with dashed lines. Panel B shows the product of the homologous recombination events. As in Figure 5.7, the linear fragment containing the *neo* gene has been substituted into the chromosome for the corresponding endogenous sequences. The replacement event does not result in the duplication of *hprt* sequences. The transcripts expected from this structure are indicated at the bottom. As with the insertion vector shown in Figure 5.13, one transcript should be initiated at the *neo* promoter and encode a functional gene product; a second transcript should be initiated at the *hprt* promoter and encode a nonfunctional Hprt gene product due to the insertion of *neo* sequences in exon 4. The overall phenotype of the targeted cell should thus be Hprt$^-$ and G418-resistant.

number of colonies per number of transfected cells. As indicated earlier, these frequencies depend on several factors, for example, the efficiency of transfection. One factor often not taken into account is the survival of cells during transfection. The mechanical stresses on cells during the transfection process can be substantial and lead to significant killing. Most important, the survival rate is dependent on the precise transfection conditions and varies widely between laboratories, even between experiments in a single laboratory. If, for example, one group reported a targeting frequency of 10^{-5} (1 targeted clone per 10^5 transfected cells) and another group a frequency of 10^{-6}, one may be led to the conclusion that the first targeting construct was more recombinogenic. Unfortunately, values such as these cannot be relied on unless the various factors affecting transfection are explicitly determined for the experiments in question.

The uncertainties in the quantitation of absolute values for recombination frequencies have led many investigators to report instead the ratio of homologous to nonhomologous recombination. This value has no units, since it is a ratio of two numbers with the same units. The perceived advantage of this method is that processes such as survival, efficiency of penetration into the cell, and so forth would affect the frequency of both homologous and nonhomologous recombination equally and would thus cancel out if the values were expressed as a ratio. The problem with this method is that the absolute frequency of recovered nonhomologous events can be profoundly influenced by the strength of the promoter used to drive the positive marker gene. The reason for this phenomenon is that, in all likelihood, not all positions in the genome are equally permissive for the expression of inserted foreign DNA. If this is true and if nonhomologous recombination is in fact a random sampling process of all possible integration sites, then a very strong promoter is likely to be adequately expressed in a greater number of positions than a weak one. Thus, the ratio of homologous to nonhomologous recombination can be profoundly changed by the choice of the vector with which the baseline frequency of nonhomologous recombination is gauged. This can especially be a problem when some genetic enrichment techniques, described in the next section, are used in the recovery of targeted events.

Special Techniques Had to Be Developed to Permit Gene Targeting in Mammalian Cells

By the mid 1980s it was generally realized that if gene targeting was ever to become a useful genetic tool in mammalian cells, something had to be done about the ratio of homologous to nonhomologous events. After all, there are so many interesting genes to target, but the disruption of only a few, such as *hprt,* can be directly selected or screened in any way. To solve the problem, two general approaches were tried, and both succeeded.

The first approach relies on physical methods. As an illustration, consider a targeting module put together from sequences of gene *X* and the *neo* marker along the lines of, for example, the *hprt* replacement construct (see Fig. 5.14). After transfection and selection in the presence of G418, approximately 10,000 colonies are recovered, the vast majority of which are due to nonhomologous events. Next, all the colonies are removed from the plates and mixed together into one large pool of cells. DNA is extracted from a fraction of the pool and physically examined for the presence of a homologous targeting event. The discrimination is based on the structure of gene *X*, since the targeted gene is clearly different, by virtue of the *neo* insertion, from its normal counterpart. Is there a method sensitive enough to positively identify a unique DNA structure present in such extremely low quantities? Until recently, the answer was no.

In 1987 a procedure designated the polymerase chain reaction, or PCR for short, was developed. The method is based on the synthesis of DNA by the enzyme DNA polymerase and the ability to arrange the replication of only one specific sequence in a complex mixture (see Fig. 5.15). The additional ability to synthesize the chosen molecule over and over again allows the PCR method to amplify a single molecule of DNA until enough of it is made to detect by conventional means. When DNA extracted from a pool of cells, as described earlier, is subjected to PCR, a single targeted event can be detected. Once a positive identification has been made that

---

FIGURE 5.15 (*Right*) Amplification of specific DNA sequences using the polymerase chain reaction (PCR). DNA is drawn in the ladder representation; the arrowheads indicate 3′ ends. The reaction mixture consists of (1) template DNA, (2) short single-stranded DNA molecules called primers, (3) DNA polymerase enzyme, and (4) nucleotide triphosphate precursors for the synthesis of DNA. Primers with two specific sequences are used. Each is complementary to one of the template strands, and their 3′ termini are oriented towards each other. The primers are synthesized chemically to be about 20 nucleotides in length. (*A*) Starting DNA molecule. (*B*) First cycle of the reaction, which typically consists of three steps. In the first step the template strands are separated by raising the temperature to a point where the hydrogen bonds holding the double helix are disrupted. In the second step the temperature is lowered to a point where the primers can anneal with the templates. The primers are added to the reaction in a vast excess over the templates, so that they anneal with the templates much faster than the two templates can anneal with each other. In the third step the temperature is raised slightly to reach the optimum for the DNA polymerase. The three steps are designated as denaturing, annealing, and elongating, with approximate temperatures of 94°C, 55°C and 70°C. The DNA polymerase is obtained from thermophillic bacteria so that it is not inactivated during the denaturing step. (*C*) Products at the end of the first cycle. (*D*) Second cycle, an exact repetition of the first cycle. (*E*) Products at the end of the second cycle. The number of template molecules doubles with each cycle: 2 at the end of first cycle, 4 at the end of second cycle, then 8, 16, 32, and so on. This is the reason the reaction is called an amplification.

at least one targeted event has occurred among the 10,000 or so non-homologous events, the experiment moves to its second stage, namely the recovery of the targeted events from the pool as clonal populations of cells. This is performed by simply subdividing the large pool into progressively smaller and smaller pools and applying PCR analysis at each stage to identify the pools that contain the targeted events. This procedure is often referred to as **sibling selection,** or **sib** selection for short (see Fig. 5.16).

Figure 5.15

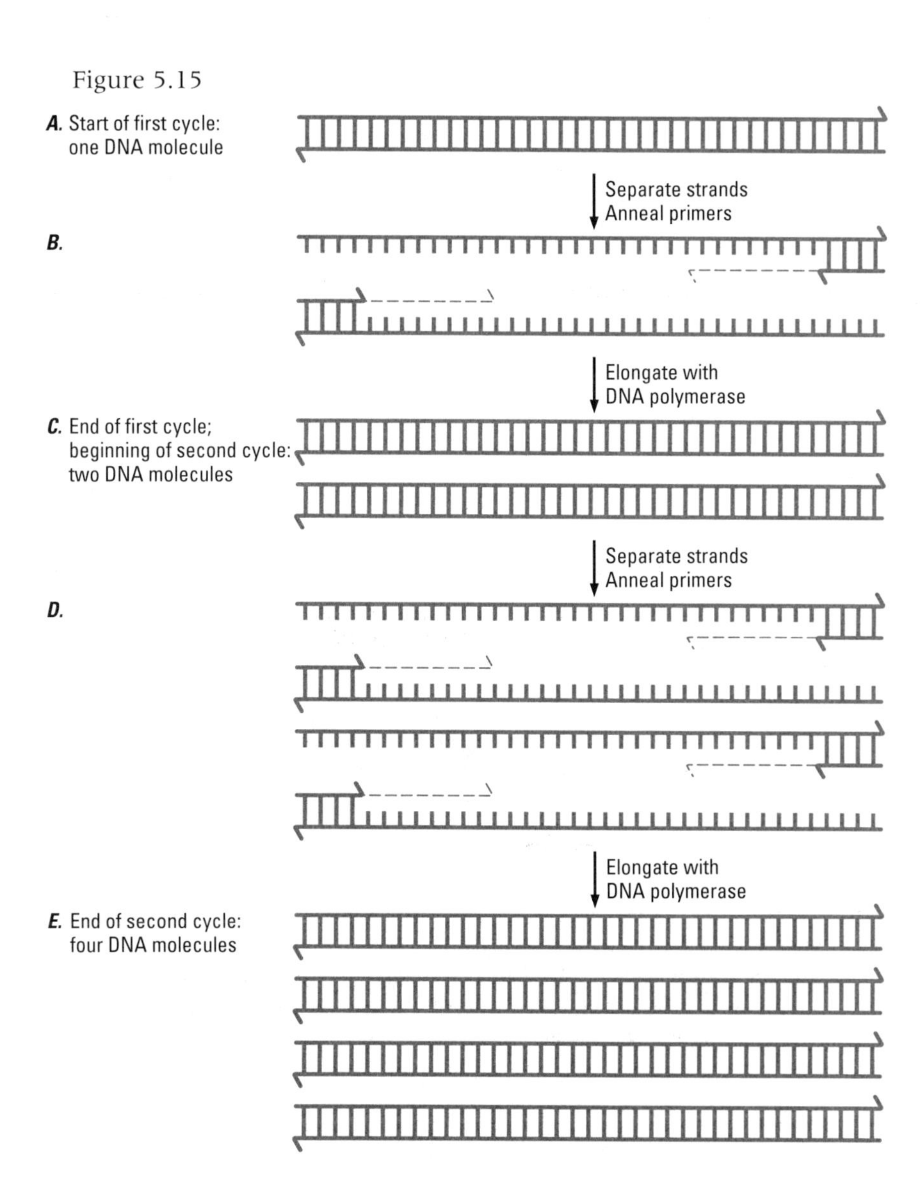

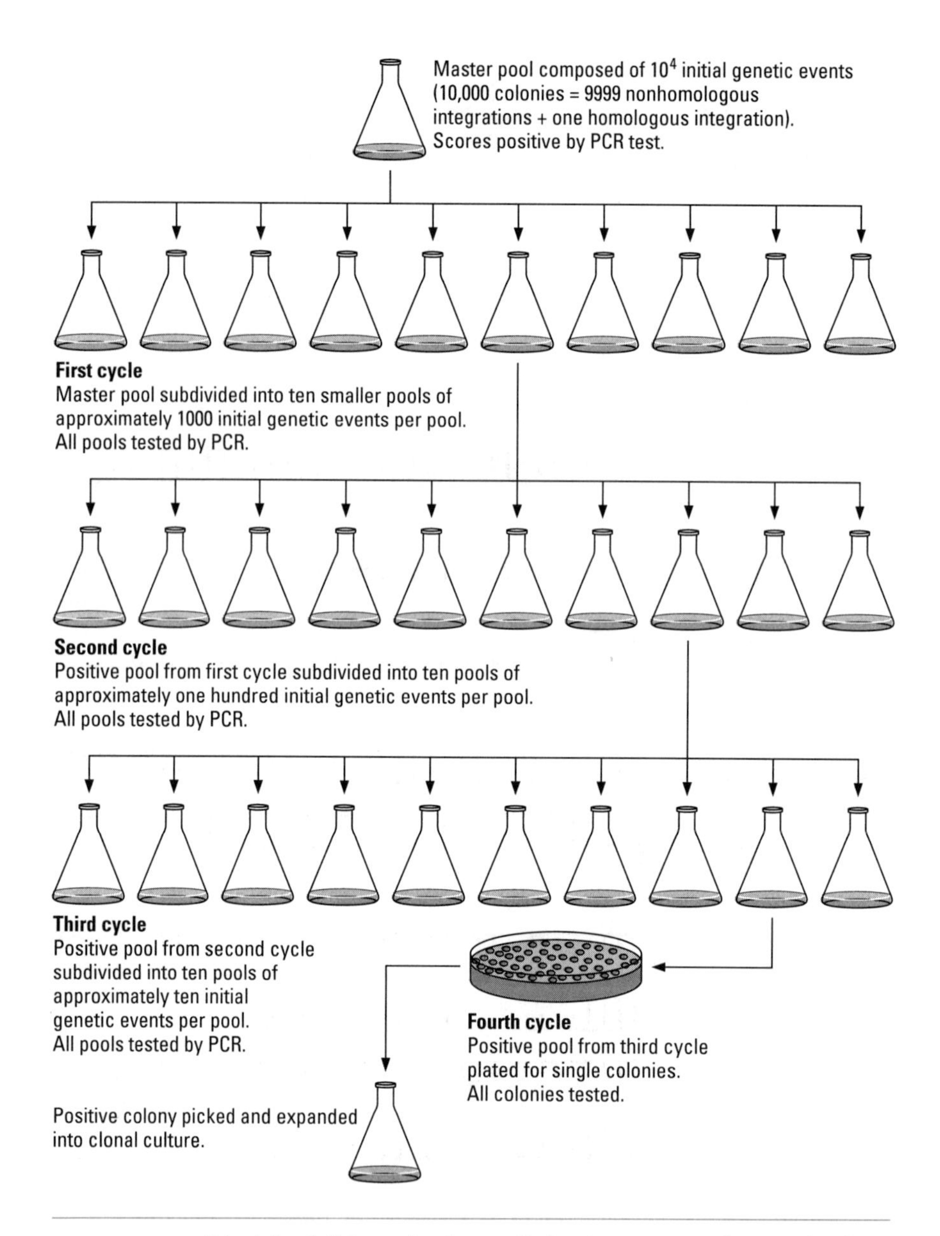

FIGURE 5.16 Principle of sibling selection applied to the recovery of targeted cells as clonal populations from pools containing a large excess of nontargeted cells. (The illustrated flasks are commonly used to grow liquid cultures of microorganisms. Although mammalian cells can be grown in a variety of differently shaped vessels, the principle of sib selection remains the same.)

In the end, the smallest pool is plated for single colonies, one of which should contain the targeted event.

The second approach relies on genetic methods. The ability to influence the absolute frequencies with which homologous and nonhomologous events occur inside cells is at the present time limited. It has been possible, however, to design genetic selections that prevent the majority of nonhomologous events from developing into colonies. Several selection methods were developed independently and reported in 1988.

The first scheme is based on the general principle of positive selection for integrations at the target locus. To achieve such selection, the positive selectable marker in a targeting module, for example, *neo* in the constructs shown in Figures 5.13 and 5.14, is modified by the removal of its promoter. For this reason the method is often referred to as **promoterless selection.** The removal of the promoter has the effect of silencing the *neo* gene by eliminating its transcription. If the *neo* gene is positioned correctly in the construct, its expression can be restored after a homologous recombination event by virtue of transcription from the promoter of the target gene. Two versions of this selection are in current use. The first one places the ATG initiation codon of the *neo* coding region at or in the 5′ direction from the ATG codon of the target gene (see Fig. 5.17). Transcription initiated at the promoter of the target locus results in the synthesis of an mRNA that contains the ATG codon of the *neo* gene in the correct context for translation. The second version of the selection places the *neo* coding region within the coding region of the target gene, such that the two translational reading frames are in register (see Fig. 5.18). Translation of the resultant mRNA initiates at the ATG codon of the target gene and produces a hybrid polypeptide chain containing the Neo sequence in its second half. Two practical restrictions must be kept in mind during the design and construction of targeting modules. First, the promoter of the target gene must not be included in the vector. Second, the target gene must be actively transcribed in the cells used in the experiment.

Following transfection, the cells are selected for resistance to G418 only; the locus to be targeted is not placed under any selection during the experiment. The principle of the selection is simply that a nonhomologous integration event should not activate the *neo* gene. The selection is not absolute, however, because a fraction of nonhomologous recombinations fortuitously integrates the construct into active cellular genes. Thus, the procedure is often appropriately referred to as an **enrichment.** Such events are relatively rare, because the integration must result in a configuration that allows not only correct transcription but also translation of the *neo* gene. The enrichment that is usually attained is approximately one hundredfold. Using an average value 10,000 : 1 for the ratio of nonhomologous to homologous recombination events of, as an example, a one hundredfold enrichment would predict a high probability of finding a

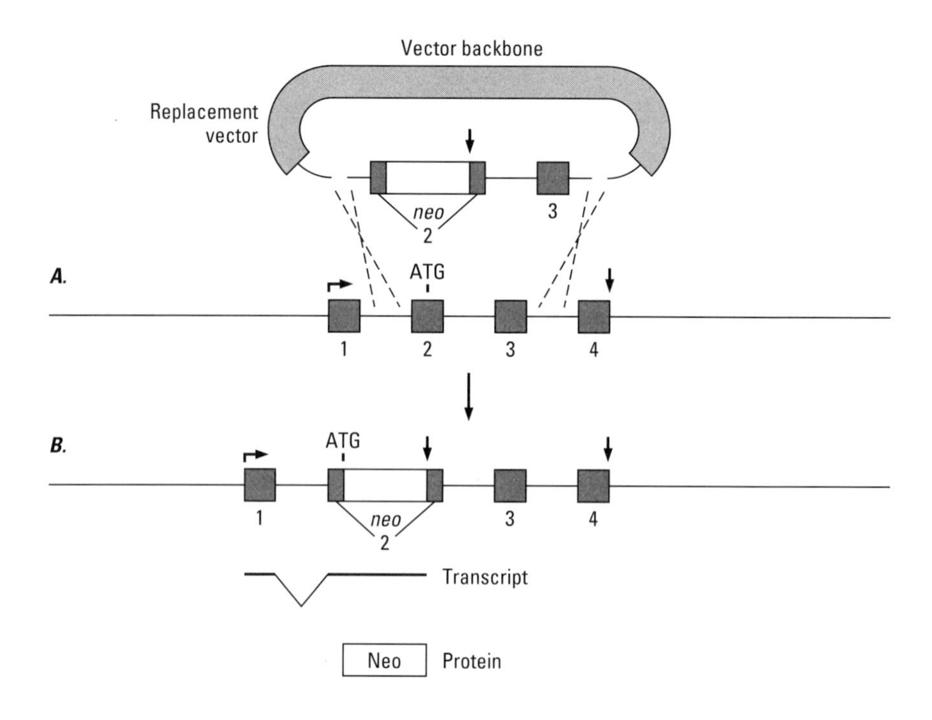

FIGURE 5.17 Positive genetic selection for gene targeting using promoter gene fusions. (*A*) Incoming targeting vector, a modified version of the replacement vector in Figure 5.14, positioned above a hypothetical chromosomal gene *X* containing 4 exons. The salient feature of the selection scheme is that the vector does not contain a promoter capable of transcribing the *neo* gene. The start of transcription of the chromosomal gene *X* is indicated with a bent arrow. Note the positions of the ATG translation initiation codons of the *neo* gene and gene *X*. In the specific example given here, *neo* is joined to gene *X* in the vector such that the fusion joint coincides with the initiation codons of both genes. Alternatively, the initiation codon of the *neo* gene could be placed anywhere in an exon 5′ of the initiation codon of gene *X*. Sites of translation termination are indicated with small downward arrows. As is usual with replacement vectors, two double-stranded breaks are introduced into the targeting vector prior to transfection. (*B*) Product of the homologous recombination event. As in Figure 5.14, the linear fragment containing the *neo* gene has been substituted into the chromosome for the corresponding endogenous sequences. In contrast to Figure 5.14, the only transcript expected of this structure should be initiated at the promoter of gene *X*, and should encode a functional Neo gene product but a nonfunctional *X* gene product due to the insertion of *neo* sequences in exon 2. The overall phenotype of the targeted cell should thus be X^- and G418-resistant. The key to the selection is that a single homologous event activates the neo gene and at the same time inactivates the chromosomal target locus. Nonhomologous recombination of the targeting vector would activate the neo gene only if the integration fortuitously occurred near a transcribed chromosomal gene. The selection scheme is not limited to replacement type vectors; it works equally well with insertion vectors.

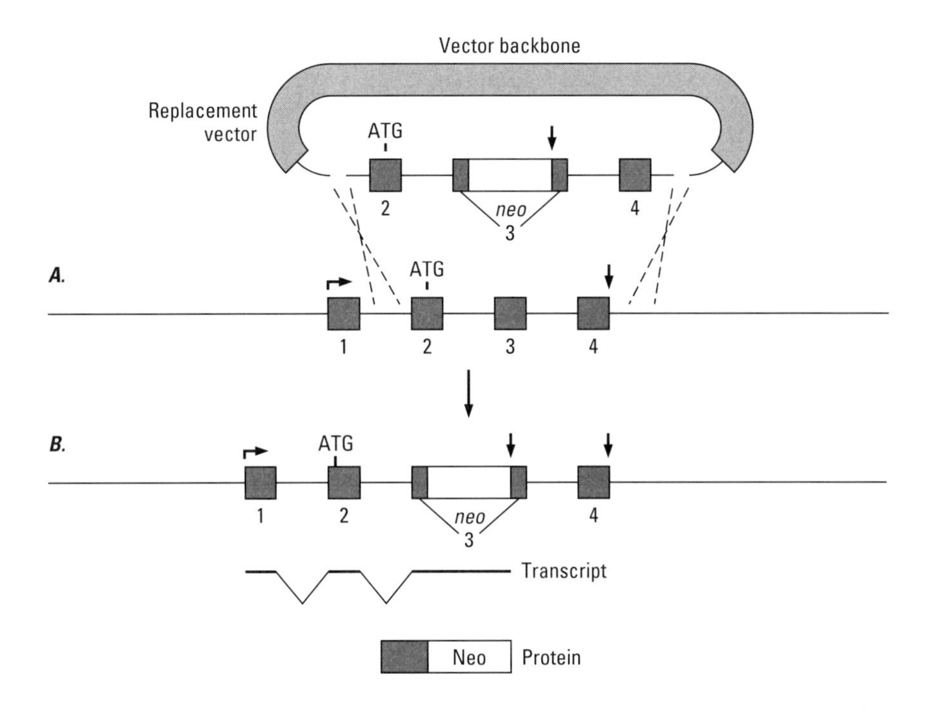

FIGURE 5.18 Positive genetic selection for gene targeting using in-frame protein gene fusions. (*A*) Incoming targeting vector, a modified version of the vector in Figure 5.17, positioned above the same hypothetical chromosomal gene *X* containing 4 exons. The symbols used are the same as those in Figure 5.17. Note the site of insertion of the *neo* gene in the vector and the positions of transcription initiation, translation initiation, and translation termination. The salient feature of this selection scheme is that the *neo* gene is joined to the target gene to produce an in-frame fusion of the two coding regions. As with the promoter fusion vector shown in Figure 5.17, this vector also lacks a promoter capable of transcribing the *neo* gene. (*B*) Product of the homologous recombination event. As in Figure 5.17, the linear fragment containing the *neo* gene has been substituted into the chromosome for the corresponding endogenous sequences. Similarly, the only transcript expected from this structure should be initiated at the promoter of gene *X*, and encode a nonfunctional *X* gene product due to the insertion of *neo* sequences in exon 3. The distinction with respect to a promoter fusion vector is in the structure of the Neo protein encoded by the targeted locus. In this case a hybrid polypeptide is produced that contains a portion of the *X* protein joined to the Neo protein. The key to the selection is that the Neo enzyme is still active, even if present as part of a larger fusion polypeptide. The overall phenotype of the targeted cell should thus be X^- and G418-resistant. Again, the selection scheme is not limited to replacement type vectors but works equally well with insertion vectors.

sought targeted clone among 100 G418-resistant colonies. In experimental terms, this is a significant achievement, since sorting through 100 or so colonies is not an unreasonable proposition.

The second genetic selection scheme is based on the general principle of negative selection against integrations at nonhomologous positions. To achieve such selection, a negative marker is included in the flank of the targeting module (see Fig. 5.19). The construct also contains the usual positive selectable marker, in this case with its own independent promoter. For this reason the method is often referred to as the **positive–negative selection,** or **PNS** for short. The method is based on the supposition that nonhomologous integrations occur primarily through the ends of a transfected molecule. In other words, the majority of linear molecules transfected into cells that go on to score as stable clones will have integrated into the genome before extensive exonucleolytic degradation had taken place. This in turn implies that a negative marker tacked onto the flank of a targeting module will be integrated intact in the majority of nonhomologous integration events. If most nonhomologous integrations preserve a flanking negative marker they can be eliminated by imposing the appropriate selection. Since the positive marker contains its own promoter, it is neutral in the selection of homologous versus nonhomologous integrations and merely serves to identify cells that have productively incorporated exogenous DNA.

In theory, there are several advantages to the PNS method of gene targeting. One is the relative ease with which targeting vectors can be constructed; since both of the markers contain their own promoters, precise junctions, for example to maintain translational frames, need not be made. Another is the absence of the requirement to exclude the promoter of the target gene from the vector, thus allowing targeting modules to have extensive 5′ and 3′ flanking regions to stimulate homologous recombination. Finally, the target gene need not be actively transcribed.

The negative marker most commonly used in the PNS selection procedure is the thymidine kinase gene, *tk* for short, from herpes simplex virus. The *tk* gene, like the *hprt* gene, is involved in what is often called a salvage pathway, in this case the utilization of pyrimidine nucleosides, namely thymidine, from the growth medium. All species, excluding some fungi, have a *tk* gene; herpes virus encodes its own to augment the cellular activity and facilitate its replication. The cellular and viral enzymes have slightly different substrate specificities, which have been exploited for the development of antiviral drugs. Such drugs, for example gancyclovir (GANC), are toxic thymidine analogs that are utilized by the viral enzyme much more efficiently than by the cellular enzyme. Thus, cells containing the viral *tk* gene will be killed in a medium containing GANC, whereas cells containing only their own *tk* gene will survive.

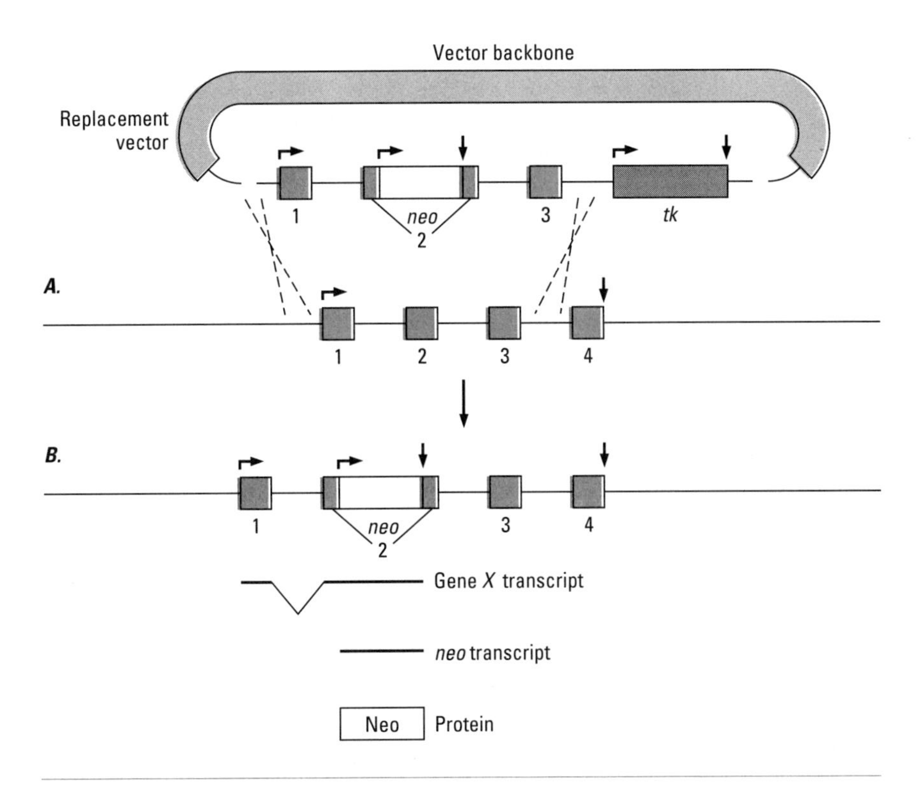

FIGURE 5.19 Positive–negative selection for gene targeting. (*A*) Incoming targeting vector positioned above the same hypothetical chromosomal gene *X* containing 4 exons. The symbols used are the same as in Figures 5.17 and 5.18. The salient feature of this selection scheme is the inclusion of a negatively selectable marker, the *tk* gene, in the vector. Another important feature is that both the *neo* and *tk* genes contain their own independent promoters. Note the sites of insertion of the *neo* and *tk* genes in the vector and the positions of transcription initiation, translation initiation, and translation termination. Prior to transfection, the vector is cut to expose one homologous end. (*B*) Product of the homologous recombination event. In particular note the position of the rightward crossover, which excludes the *tk* gene from integration in the chromosome. The end product of the recombination is thus the substitution of a linear fragment containing the *neo* gene into the chromosome, displacing the corresponding endogenous sequences. The transcripts expected from this structure are indicated at the bottom. One transcript should be initiated at the *neo* promoter and encode a functional gene product; a second transcript should be initiated at the gene X promoter and encode a nonfunctional *X* gene product due to the insertion of *neo* sequences in exon 2. The overall phenotype of the targeted cell should thus be X^- and G418-resistant.

Following transfection, the cells are selected for resistance to both G418 and GANC; however, the locus to be targeted is not under selection. The principle of the selection is simply that a nonhomologous integration will incorporate the *tk* marker into the genome and will not be allowed to score as a colony in the presence of GANC. This selection is, however, also not absolute, since damage suffered by the transfected DNA can inactivate the negative marker and thus eliminate the selection. The frequency of such damage-induced inactivations probably varies from cell type to cell type and may also be influenced by the specific transfection protocol. In any case, this procedure also ends up being merely an enrichment. The enrichment ratios that have been reported are quite variable: anywhere from one thousandfold all the way down to less than twofold. Nevertheless, the method has received widespread use and numerous loci have been successfully targeted with it.

Gene Targeting in Mammalian Cells: Whither Now?

The emergence of gene targeting as a practical method has had a profound impact on genetic analysis. The majority of gene targeting studies to date have been performed in mouse embryonic stem cells; these rather special cells can be regenerated into whole, live animals. This application of gene targeting is treated in detail in the following chapters. Somatic tissue culture cells have also been used, but they present some special problems. The first problem is that in a diploid cell both copies of a gene have to be disrupted before the loss-of-function phenotype can be manifested. The second problem is that if a gene is required for viability, disruption of the second copy would lead to cell death.

The first problem can be circumvented simply by performing two successive rounds of gene targeting. As gene targeting procedures become more and more streamlined, such a goal becomes less and less formidable. It has in fact been accomplished in a few cases. The second problem can be circumvented by placing the experimental gene under conditional control. The most widely used method towards this end is the use of promoters whose activities can be regulated at will by extracellular stimuli. Some promoters, for example, can be regulated by hormones. In the absence of the specific hormone the promoter is off, and addition of the hormone to the medium triggers transcription. One possible scheme to target an essential gene *X* could be to (1) construct a vector containing a copy of gene *X* driven by a regulatable promoter, (2) integrate the vector into the genome by nonhomologous recombination, (3) ascertain that the engineered gene *X* can be adequately regulated by the appropriate extracellular stimuli, and (4) perform two gene targeting knockouts on the natural copies of gene *X*. A cell line such as this would be expected to remain viable as long as the introduced conditional copy of gene *X* were kept active; turning off the regulatable promoter would then elicit the loss-of-function phenotype. Multistep genetic manipulations of this complexity are just beginning to be attempted in mammalian cells.

An even more sophisticated scheme would be to introduce a conditional promoter directly into one (or both) of the natural, endogenous copies of gene *X*. Such a manipulation falls well beyond the scope of simple gene disruptions and necessitates the use of methods such as the pop-in/pop-out or direct gene modification procedures. Some limited success has already been achieved with such fine gene targeting manipulations in mammalian cells, but these methods still need to be perfected before they can be as readily applied as simple gene disruptions.

Suggested Readings

Bollag, R. J., Waldman, A. S., and Liskay, R. M. (1989). Homologous recombination in mammalian cells. *Annu. Rev. Genet.* 23: 199–225.

Capecchi, M. R. (1989). Altering the genome by homologous recombination. *Science* 244: 1288–1292.

Evans, M. J. (1989). Potential for genetic manipulation of mammals. *Mol. Biol. Med.* 6: 557–565.

6

TRANSGENIC MICE—TWO APPROACHES

A **transgenic** organism is one that has an extra or exogenous fragment of DNA in its genome. Various methods have been developed for making transgenic plants as well as some species of animals, with the most common research animals being the nematode worm, the fruit fly, and the mouse. In order to achieve stable inheritance of the exogenous DNA fragment, the integration event must occur in a cell type that can give rise to functional germ cells, either sperm or oocytes. Two mouse cell types that can form germ cells and into which DNA can be readily introduced are fertilized egg cells and embryonic stem cells.

At present, mouse embryonic stem cells, commonly referred to as **ES cells,** are preferred for gene targeting experiments since (1) they can be screened in culture for rare homologous recombination events, (2) homologous recombination with some vectors occurs in these cells at a high frequency, and (3) the cells can be returned from in vitro culture to a "host" embryo where they become incorporated into the developing mouse. Of greatest importance, the mouse embryonic stem cells can give rise to cells of all tissues, including germ cells. To create a targeted mutation in mice, the homologous recombination event is performed in embryonic stem cells in culture and then the mutation is transmitted into the germline by injecting the cells into an embryo. The mice carrying mutated germ cells are then bred to produce transgenic offspring that are heterozygous for the mutation.

Zygote Injection

A number of techniques are available for making transgenic mice, the most efficient of which is zygote injection. This method involves injecting DNA into a fertilized egg, or **zygote,** and then allowing the egg to develop in a pseudopregnant (see below) mother (Figs. 6.1 and 6.2). The transgenic animal that is born is called a **founder,** and it is bred to produce more animals with the same DNA insertion. In this method of making transgenic animals, the new DNA typically randomly integrates into the genome by a nonhomologous recombination event (Fig. 6.3). One to many thousands of copies of the DNA may integrate at one site in the genome. Homologous recombination has been found to occur in somatic cells in culture after delivering the DNA to the nucleus by injection. It therefore should be feasible for germline gene targeting to be achieved by zygote injection. This, of course, requires that zygotes undergo homologous recombination at a high frequency. The results to date have

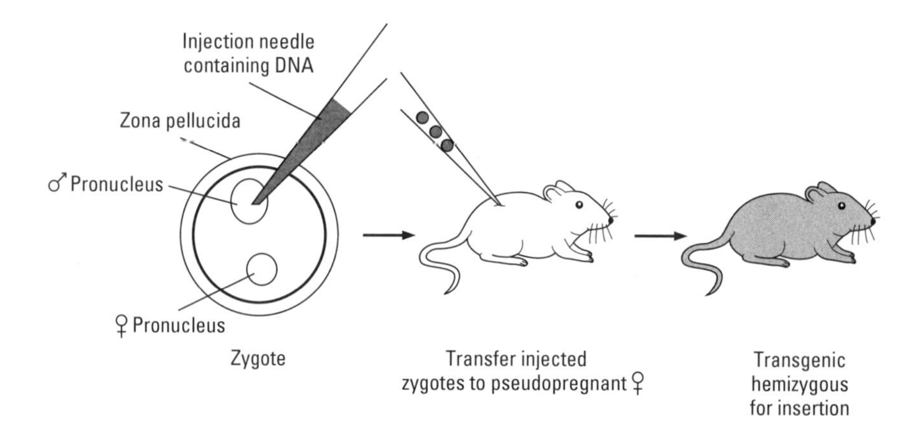

FIGURE 6.1 Procedure for making transgenic mice using zygote injection. Zygotes are removed from the oviducts of a female mouse that has mated the night before. The zygotes must be released from the cumulus cells that surround them. The zygotes are placed under a microscope with two micromanipulator setups. The zygote is held in place with a blunt holding pipette and the tip of the injection needle is filled with a solution containing DNA (~5 μg/ml). The DNA is then injected into one of the pronuclei, usually the larger male pronucleus. The zygotes are then either transferred the same day, or cultured overnight to form 2-cell embryos and then transferred, into the oviducts of 0.5-day pseudopregnant females. Approximately 50 percent of the eggs survive to the 2-cell stage and approximately 20 percent to term. Twenty to thirty embryos are therefore transferred to the uterus of each female. Of the animals born, 10 to 30 percent should be transgenic (hemizygous) and contain one site of DNA integration.

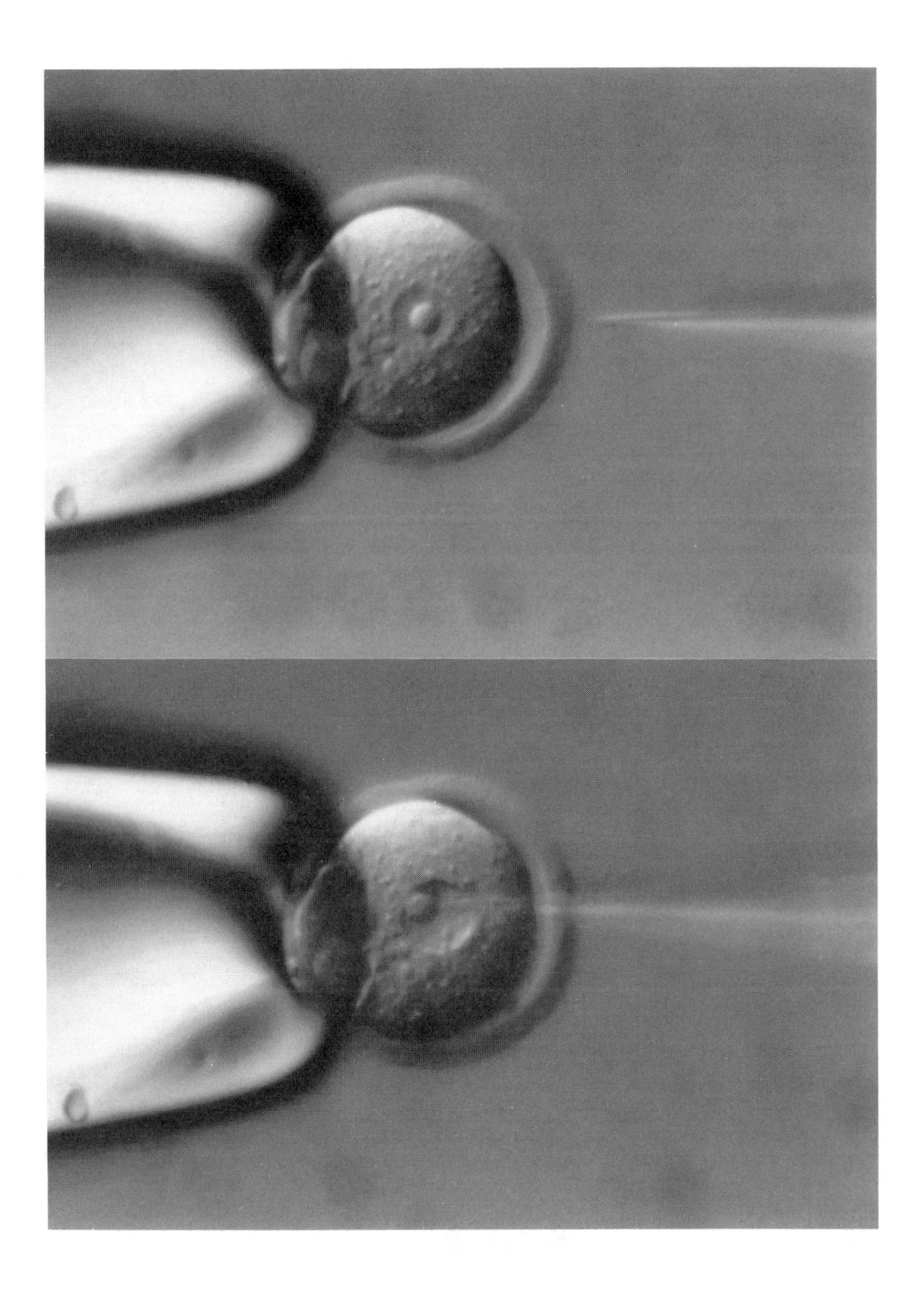

FIGURE 6.2 Microinjection of DNA into a mouse zygote. *Top*: The holding pipette (left) is keeping the zygote (middle) in place while the injection pipette containing DNA (right) is brought towards the egg. The nucleus can be seen in the middle of the egg. *Bottom*: The injection needle is inside the nucleus of the zygote. The nucleus has expanded in size due to the injection of DNA. (*Photographs courtesy of Anna Auerbach.*)

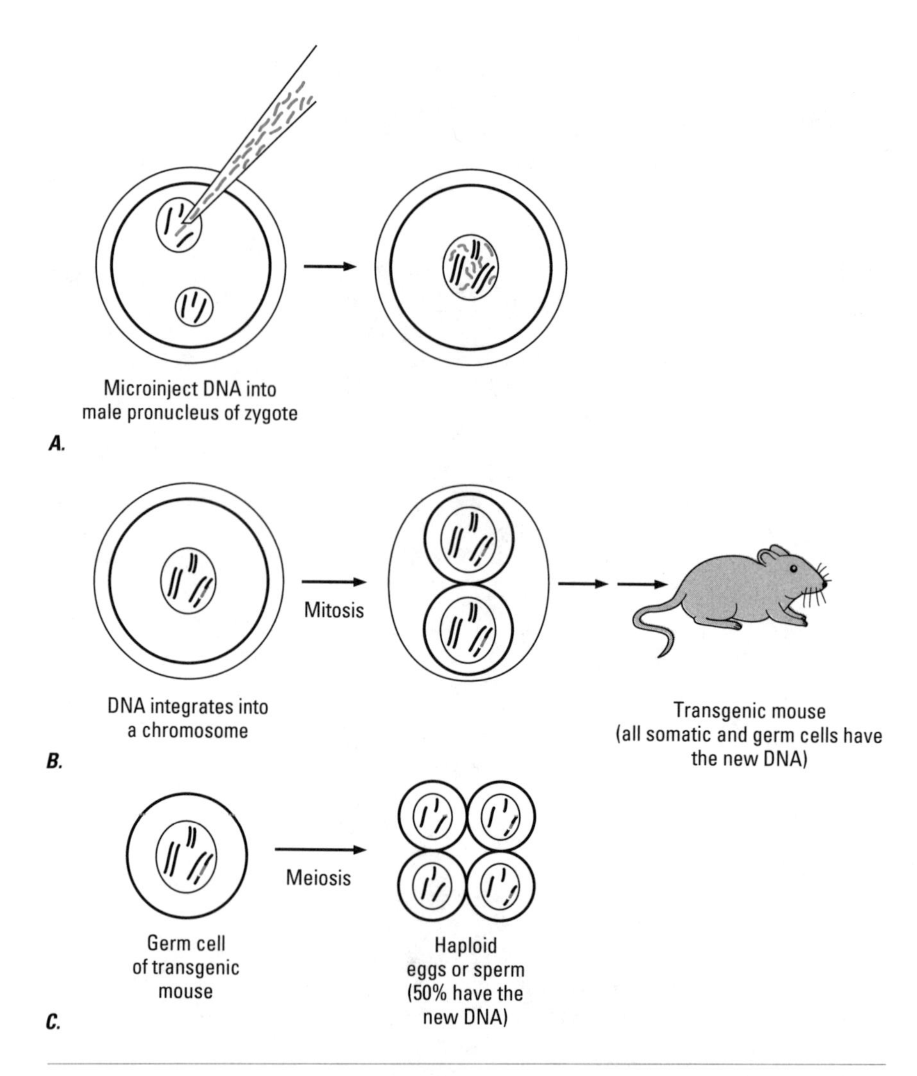

FIGURE 6.3 Random integration of DNA at one site in a transgenic. Three pairs of chromosomes are shown as pairs of lines of different lengths. The integrated DNA is shown as a green line. (*A*) DNA is injected into a pronucleus of a zygote and the two pronuclei fuse. (*B*) The DNA inserts randomly at one site in the genome of the zygote. The egg develops and all somatic cells of the transgenic contain one copy of the inserted DNA. (*C*) A germ cell from the transgenic is shown schematically. Following meiosis, only two of the four haploid germ cells contain the transgene insertion.

not been encouraging, although further investigation is required in which the same targeting vector is compared in ES cells and zygotes.

Regardless of such results, the big advantage of ES cells over zygote injection for germline targeting is that it is relatively easy to screen 1000 ES cell clones (see Fig. 7.9) compared to making and screening 1000 transgenic animals. Since the frequency of homologous recombination can vary many orders of magnitude from locus to locus, zygote injection is likely to be limited to genes that have been tested in ES cells and shown to give a high frequency of homologous recombination relative to non-homologous integration.

Early Embryonic Development

An understanding of the basic developmental events that occur during the first nine days of mouse embryogenesis is necessary for a full comprehension of mouse embryonic stem cells. The function of embryonic development is to produce from a single egg a three-dimensional organism with a body plan that is predetermined by the genetic code. The process of embryogenesis involves a progressive loss of the developmental potential of cells. The fertilized egg has the capacity to give rise to all cell types in the adult mouse and is thus said to be **totipotent**. As the embryo cells divide, however, some lose their ability to form all cell types and retain only a restricted capacity to develop into certain tissues. A cell that has undergone this process of becoming restricted is said to have **differentiated**. Development involves many differentiation events that occur in series and in parallel. An example of this is shown in Figure 6.4.

Embryonic development begins with fertilization of the egg and the subsequent joining of the two haploid nuclei from the mother and the father to form a diploid zygote. The egg is surrounded by a layer of mucopolysaccharides called the **zona pellucida** and has a diameter of approximately 100 μm. The first four days of development (Fig. 6.5) occur in the oviducts and uterus before the egg becomes implanted in the wall of the uterus. During this preimplantation stage no growth of the embryo occurs, since there is no external source of nutrition. After implantation, however, rapid growth is possible since the embryo makes direct contact with the mother's blood supply through the development of a set of specialized extraembryonic tissues that form the embryonic component of the placenta and the membranes that surround the embryo.

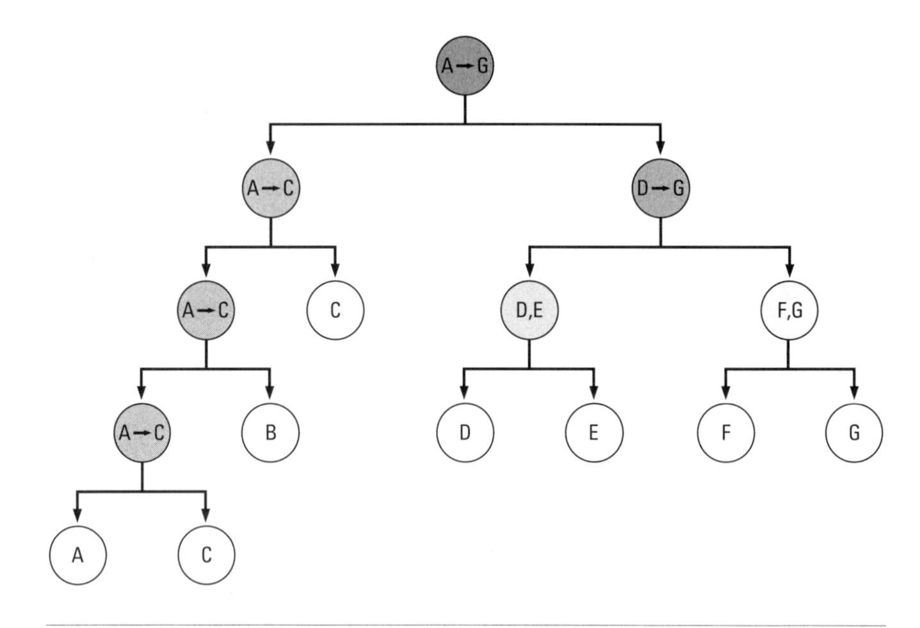

FIGURE 6.4 Lineage of a multipotent cell (A→G). Cell A→G divides and gives rise to two daughter cells (A→C; D→G) that have more restricted and different developmental potentials. Cell A→C divides three times and gives rise to three differentiated cell types (A; B; C). Cell D→G divides only once to give rise to two cells (D, E; F, G) with more restricted developmental potential. These two cells divide once each to give rise to four different differentiated cell types (D; E; F; G)

The first two days of development are referred to as the early cleavage stage. The embryo undergoes three cell divisions during this period (Fig. 6.5*B, C, D*) to form an 8-cell embryo, called a **morula**, in which all the cells appear to be totipotent and thus have an equivalent developmental potential. During the next 24 hours, however, the embryo undergoes two more rounds of cell division, the cells compact, and the ones on the outside of the morula differentiate to form a layer of cells called **trophectoderm**. The undifferentiated totipotent cells on the inside are called the **inner cell mass (ICM) cells** (Fig. 6.5*E*). A fluid-filled cavity, or blastocoel, forms in the interior, leaving the ICM as a clump of cells attached to one end of the outer layer of trophectoderm (see Fig. 6.5*F*). This approximately 32-cell embryo is referred to as a **blastocyst**. It is the ICM cells that, when transferred into culture, give rise to embryonic stem cell lines.

The trophectoderm goes on to form extraembryonic tissues, whereas the ICM gives rise to additional extraembryonic tissues as well as to all the

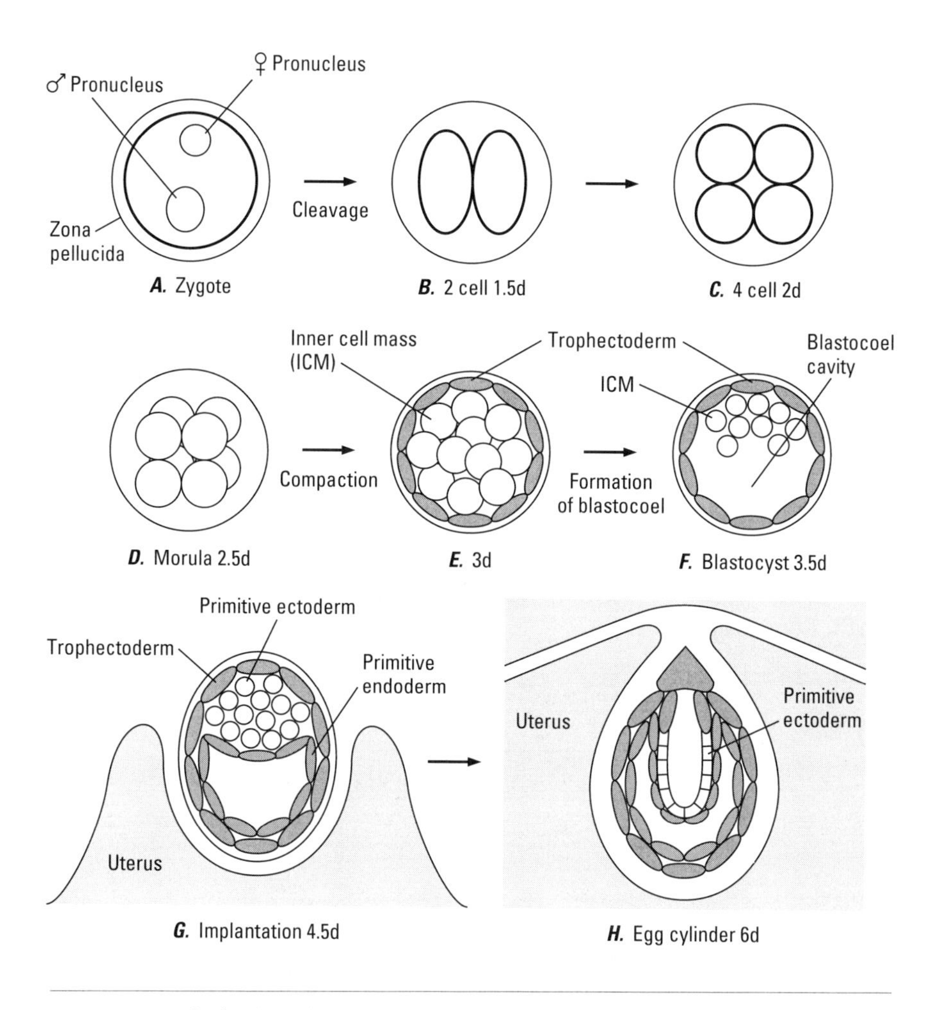

FIGURE 6.5 Preimplantation and early postimplantation development. The days of development indicated are postfertilization. Diagrams *E–H* depict a representative view of what the middle of an embryo (sagittal section) would look like if the embryo were cut in half. Trophectoderm cells and their derivatives are shown in dark green, primitive endoderm in black, the cells that will give rise to the embryo (zygote up to morula, inner cell mass, and primitive ectoderm) in white, and the uterus in light green.

cells of the embryo. During the next day the blastocyst fully expands, "hatches" from the zona pellucida, and becomes implanted in the uterus. The uterus of the female is competent to receive the embryo for only about five days after mating with a male. A female in estrous and mated with a vasectomized male is referred to as a **pseudopregnant** mother since she is competent to receive embryos but does not contain any fertilized eggs. Pseudopregnant mothers are important for making transgenic mice since they serve as the surrogate mothers for embryos that have been injected with DNA or embryonic stem cells.

During implantation, the cells of the ICM facing the blastocoel differentiate into a layer of **primitive endoderm** cells that migrate and cover the inner wall of the trophectoderm. These cells also give rise to extraembryonic tissues. The inner cells that will give rise to the embryo are referred to as the **primitive ectoderm** (Fig. 6.5*G*). The trophectoderm cells adjacent to the primitive ectoderm proliferate and push a fingerlike projection including the primitive ectoderm down into the blastocoel. A cavity then forms in the projection and the primitive ectoderm takes on the form of a cylinder that is a single cell layer thick (Fig. 6.5*H*). Up until this **egg cylinder** stage of development (six days), the cells that will give rise to the embryo remain undifferentiated and are set aside first as the ICM and then as the primitive ectoderm. The only development of defined or differentiated cell types involves the extraembryonic tissues.

Gastrulation and Development of the Embryo

During the next three days of embryogenesis (6.5 to 9.5 days), the primitive ectoderm undergoes a complicated set of events involving regional differences in cell proliferation, precise cell movements, and differentiation of the three cell layers to form an embryo that has the basic features of a mouse. This stage of embryogenesis, called **gastrulation,** is perhaps the most complicated and critical period of development. At 6.5 days, a group of cells in the primitive ectoderm adjacent to the trophectoderm begin to separate from the surrounding cells and move under and spread out across the ectoderm. The cells invaginating through the ectoderm form a sort of trough that is called the **primitive streak** (Fig. 6.6*A*). It is not known what stimulates a group of cells to begin gastrulation, nor is it possible to predict which cells in the egg cylinder will form the primitive streak. The position where the cells begin this process marks the future posterior end of the embryo and defines where the midline of the animal will be. During the next day the primitive streak extends toward the anterior end of the embryo (Fig. 6.6*B*). These cells lay down a layer of mesoderm and under it a layer of endoderm. One of the functions of gastrulation is therefore to differentiate the three germ layers of the

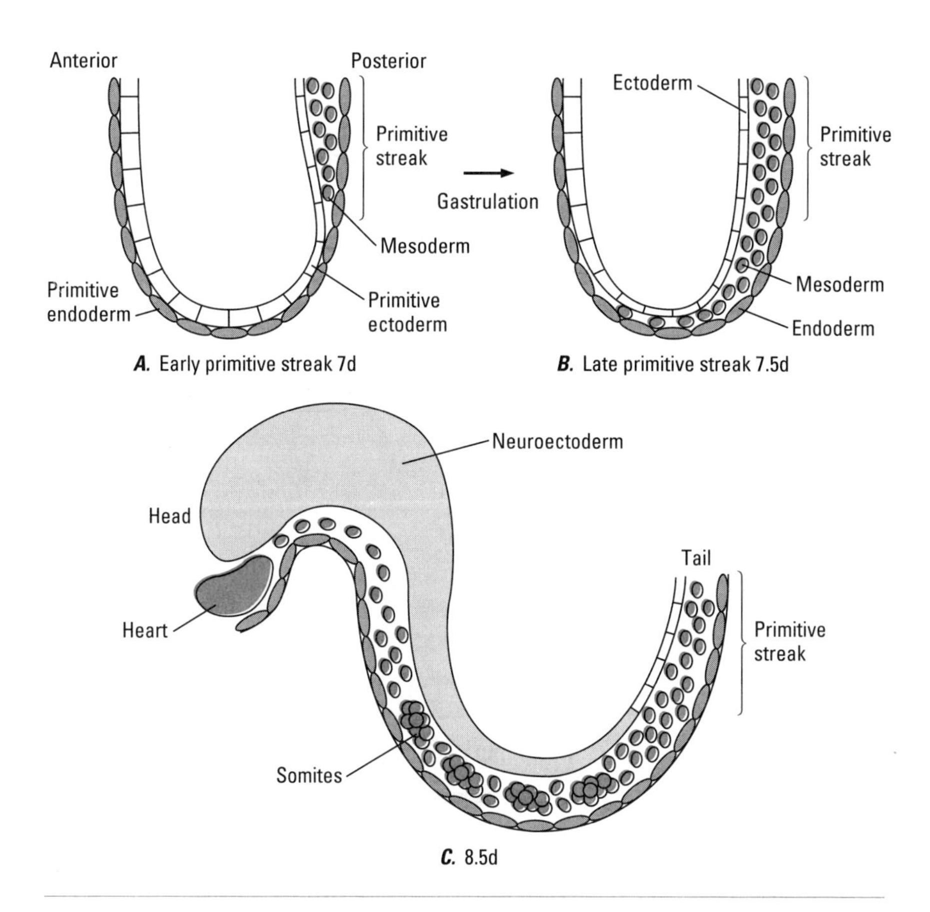

FIGURE 6.6 Gastrula-stage embryos. Representative sagittal sections through an embryo are shown, as in Figure 6.5*E–H*. The extraembryonic tissues, except primitive endoderm, are not shown. The days of development are postfertilization. The primitive ectoderm is shown in white, embryonic mesoderm in green, and endoderm as grey. The primitive endoderm is replaced during gastrulation by embryonic endoderm from ectoderm cells that move through the primitive streak and differentiate. The neuroectoderm (precursor of the central nervous system) that is derived from midline ectoderm is shown in grey.

embryo: ectoderm, mesoderm, and endoderm. Another important function is to establish the three basic body axes: anterior/posterior (head/tail), dorsal/ventral (back/front), and right/left sides. This is the beginning of the process of **pattern formation,** or laying down the body plan. The three germ layers then begin to subdivide to give rise to particular tissues. For example, the ectoderm in the middle of the embryo develops into **neuroectoderm** that later gives rise to the entire central nervous system.

(Fig. 6.6*C*). A series of groups of mesodermal cells lying on either side of the midline form epithelial balls called **somites** that later give rise to the vertebrae, the muscles of the trunk and limbs, and the dermal (inner) layer of the skin. Endoderm cells form the inner layer of the digestive tract, lungs, and liver.

The next stage of pattern formation involves regional development, or specialization within tissues. This process requires that cells at different positions along seemingly homogeneous tissues like the neuroectoderm or along the series of repeated somites later develop into the cell types appropriate for their positions along the body. For example, the anterior end of the neuroectoderm must form the brain, whereas posterior regions form the spinal cord. By 12.5 days of embryonic development, the precursors, or primordia, for all of the organs of the adult mouse have been formed. The last seven days of embryonic development involves extensive growth of the embryo and further development and differentiation of the organs in a process called **organogenesis.**

Formation of Chimeras

From the description of embryonic development in the previous sections, it can be seen that embryogenesis involves a number of events. During the preimplantation and early postimplantation period, differentiation of the extraembryonic tissues occurs, whereas the cells that will give rise to the whole embryo are temporarily set aside and proliferate without differentiation. The next event, gastrulation, involves rapid cell proliferation and cell movement and the establishment of the basic body plan. Finally, with all the body structures in place, the organs grow in size and the cells further differentiate to their final cell types. Before gastrulation, there is extensive cell mixing within the ICM and primitive ectoderm. Thus, if one ICM cell is marked at the blastocyst stage such that all of its progeny can be identified after gastrulation, the cells of this clone would likely be found in all tissues and be highly intermixed with nonmarked cells.

This extensive cell mixing works to advantage when it comes to making chimeric animals. A **chimera** is an animal that is made up of cells from two different embryos. Such animals are made by mixing together two morula-stage embryos or by injecting ICM cells from one embryo into another "host" blastocyst (see Fig. 6.7). The embryos are then transferred

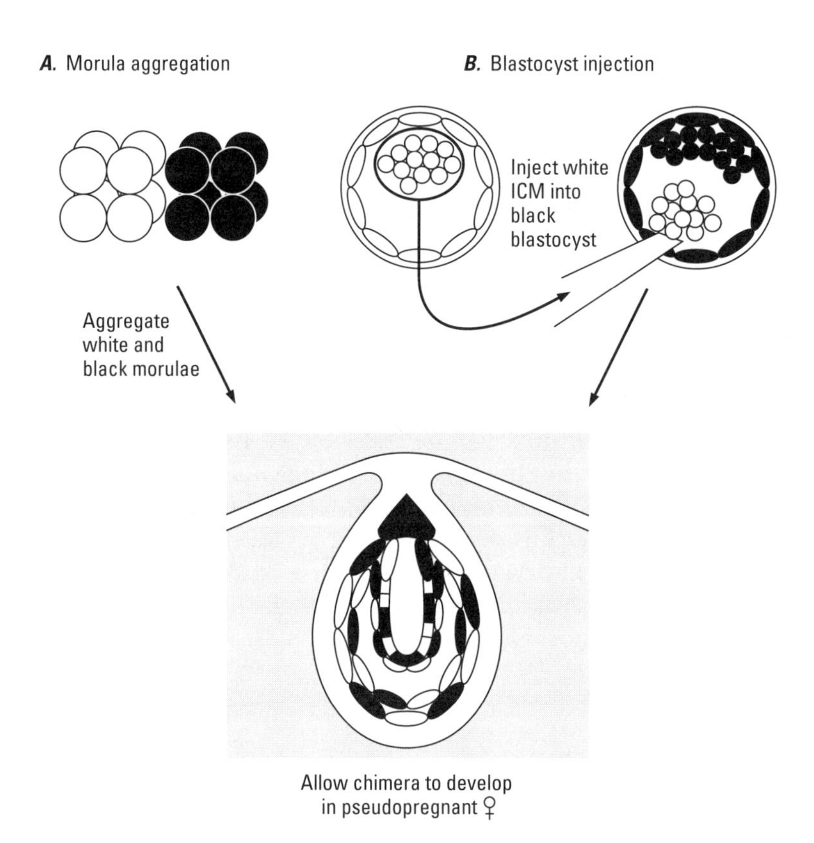

FIGURE 6.7 Two procedures for making chimeric embryos. One embryo is shown in black and the other in white to indicate two different genotypes. At the bottom is shown a sagittal section through a 6-day chimeric embryo in which cells from each embryo (black and white) are randomly mixed in all cell layers. (*A*) Morula aggregation: To obtain morulae, the oviducts are removed from 2.5-day pregnant females and flushed with medium to wash out the embryos. The zona pellucida is removed from the morulae. Two morulae, each from a different genetically marked mother, are then placed together in a small depression on a dish. Gravity forces the embryos together and the cells intermix and develop into a blastocyst during the following 24 hours of incubation. The blastocysts are then transferred to the uterus of a 2.5-day pseudopregnant female (approximately 12 embryos/uterus). Approximately 60 percent of the embryos should be born at term (17 days later) and approximately 70 percent of these should be chimeras. (*B*) Blastocyst injection: Blastocysts are flushed from the uteri of 3.5-day pregnant females of two genotypes. The inner cell masses (ICM) are isolated from the blastocysts of one genotype by immunosurgery that selectively kills the outer trophectoderm cells. The ICMs and blastocysts are placed under a microscope with two micromanipulator setups. One micromanipulator is used to hold the blastocyst in place with the holding pipette. The other micromanipulator is used to hold the ICM and inject it through a tiny hole made in the blastocyst. The injected blastocysts are then transferred to the uterus of a 2.5-day pseudopregnant mother. Approximately 80 percent of the embryos should reach term and approximately 90 percent of these be chimeric.

into the uterus of a pseudopregnant mother where they implant and continue embryonic development. The resulting chimeric animals that are born are composed of a mixture of cells from both embryos in most, and in many cases all, tissues.

To test whether the germ cells are derived from either or both embryos, each embryo is marked with a genetic trait that can be recognized in the chimera and in its offspring. The most common genetic marker used is coat color. For example, if ICM cells from a white mouse homozygous for a recessive mutation at the albino locus (c/c) are injected into a blastocyst from a black mouse homozygous for the normal, or wild type, albino gene (C/C), then any resulting chimeras can be identified by the presence of patches of white and black fur. To test whether the germlines of the chimeras are made up of cells from either or both genotypes, the chimeras are bred with a white mouse (c/c). Since black is dominant over white coat color, any white offspring must be homozygous c/c and thus be derived from the injected ICM cells, whereas any black mice must be heterozygous C/c and be derived from the host blastocyst (see Fig. 6.8).

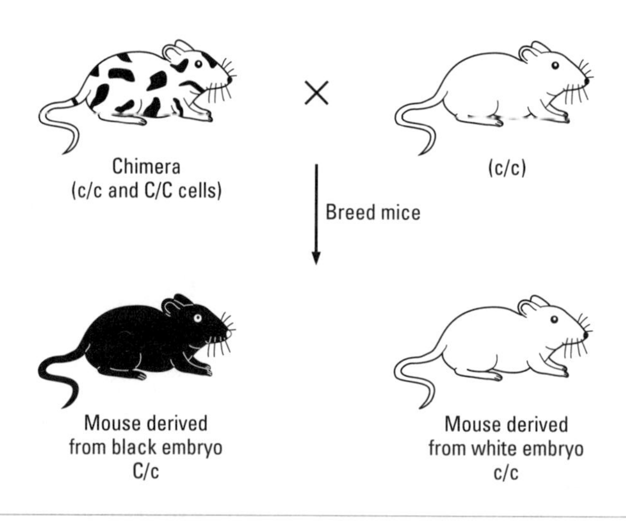

FIGURE 6.8 Breeding scheme for determining the genotypes of the cells that have populated the germline. The chimera is black and white, indicating that the genotypes of the two embryos used to form the chimera were homozygous wild type for albino (C/C), giving a black coat color phenotype, and homozygous mutant for albino (c/c), giving a white coat color. The wild type allele is dominant, and thus offspring of a chimera bred with a white (c/c) mouse can be either black and heterozygous for albino (C/c) or white and homozygous mutant for albino (c/c). The C/c mice are derived from the C/C morula and the c/c mice from the c/c morula.

Embryonic Stem Cells

In 1981, two independent groups demonstrated that ICM cells from a mouse blastocyst can be propagated in culture[1] (see Fig. 6.9). This new type of cell line is the ES cell line. One distinct property of ES cells is that they remain diploid even after being cultured for many weeks. This is in contrast to other tissue culture cell lines that often do not remain diploid but spontaneously gain and lose chromosomes at a high rate. A second unique property of ES cells is that they remain totipotent and maintain the ability, like ICM cells, to form chimeras. These two properties, maintaining a normal karyotype and extensive contribution in chimeras, are both necessary for ES cells to form functional germ cells in chimeras. ES cells can be used to make transgenics by introducing a DNA vector into them and then forming and breeding chimeras (see Fig. 6.10)

In order to maintain ES cells that can contribute to the germ cells, they must be grown in conditions that inhibit differentiation and instead promote proliferation and retention of a diploid number of chromosomes. Many different factors probably determine whether ES cells remain pluripotent. The first critical factor that was identified as a requirement for proliferation of ES cells was a layer of fibroblast cells seeded under the ES cells. The fibroblast cells are treated before seeding with irradiation or drugs to inhibit further cell division but not metabolism. These cell layers, referred to as **feeder** cells, produce and excrete proteins, one or some of which are essential for ES cell proliferation. In the absence of this cell layer, ES cells divide only a few times and differentiate into cell types typical of an early embryo, usually primitive endoderm, that are unable to further divide. A recently identified protein factor, called either Differentiation Inhibiting Activity (DIA) or Leukemia Inhibiting Factor (LIF), was found to allow ES cells to proliferate in the absence of feeders. DIA/LIF is an interesting factor since depending on the cell type, ES or leukemia cells, the cellular response to the factor is either to proliferate or to differentiate. Some laboratories are now growing ES cells in LIF in the absence of feeders. This is much simpler than preparing feeder cells, but there is some evidence that LIF, unlike feeder cells, does not completely satisfy all the growth requirements of ES cells.

The second factor that influences the pluripotential status of ES cells is the length of time they are grown in culture. With most ES cell lines there is an inverse correlation between the number of times the cells are passaged and the percentage of chimeras made with them that have an ES cell contribution to the germline. This is because abnormal variants accumulate in the cell population. To minimize the number of cell passages for each cell line, newly established ES cell lines are expanded immediately to produce a large number of cells and the cells are then frozen in small portions for future use. This produces a large stock of cells, all frozen after a minimum of passages, that can be used for years to come. Typically, an aliquot is thawed, grown, and used for experiments for four to six weeks and then discarded. A new frozen sample can then be thawed. Because of the presence of variants in the cell line population, when cells are subcloned, or isolated as single ES cells and grown up as clones, many of the clones will be derived from the variants that cannot populate the germline. A stringent test of the state of an ES cell line and the growth conditions used, therefore, is to subclone several lines and test how many of them can contribute to the germline. Ideally, at least 50 percent should remain totipotent.

A third factor that probably affects the status of ES cells is the way in which they are handled day to day by an investigator. This is obviously much more difficult to standardize than a growth factor or number of passages. However, many different labs have now successfully targeted a gene in ES cells and had the cells contribute to the germline in chimeras.

Since ES cells are derived from the ICM cells of a blastocyst and the ICM gives rise to the whole embryo, it is not surprising that ES cells also have the capacity to give rise to the whole embryo. To accomplish this, diploid ES cells, like ICM cells, are either aggregated with two morulae or injected

FIGURE 6.9 (*Right*) Procedure for establishing embryonic stem cell lines and differentiating them in vitro. Blastocysts are isolated as described in the legend for Figure 6.7. Either the blastocyst is placed directly in culture, or the ICM is removed, as described in the legend to Figure 6.7, and placed in culture. The culture medium must include LIF if a feeder layer is not provided. Feeder layers are prepared by treating either STO fibroblast cells or primary embryo fibroblast cells from a 15-day embryo with irradiation or mitomycin-C to inhibit further cell divisions. Approximately 25 percent of the ICMs can proliferate in culture and produce cell lines. ES cell lines are passaged at a high density every two to three days to avoid differentiation and selection for variants that cannot form germline chimeras or teratocarcinomas. To stimulate embryoid body formation, the cells are seeded at a high density in a bacteriological Petri dish. The cells readily aggregate, and within four to six days the outer layer of cells differentiates to primitive endoderm. These structures are called embryoid bodies. During the next two weeks the embryoid bodies expand, forming a cavity on the inside, and some ectoderm cells form mesoderm. After about twenty-one days, the fully expanded embryoid bodies can be placed in a tissue culture dish to which they attach. During the following two weeks the cells proliferate, migrate across the dish, and differentiate into cell types derived from the three germ layers.

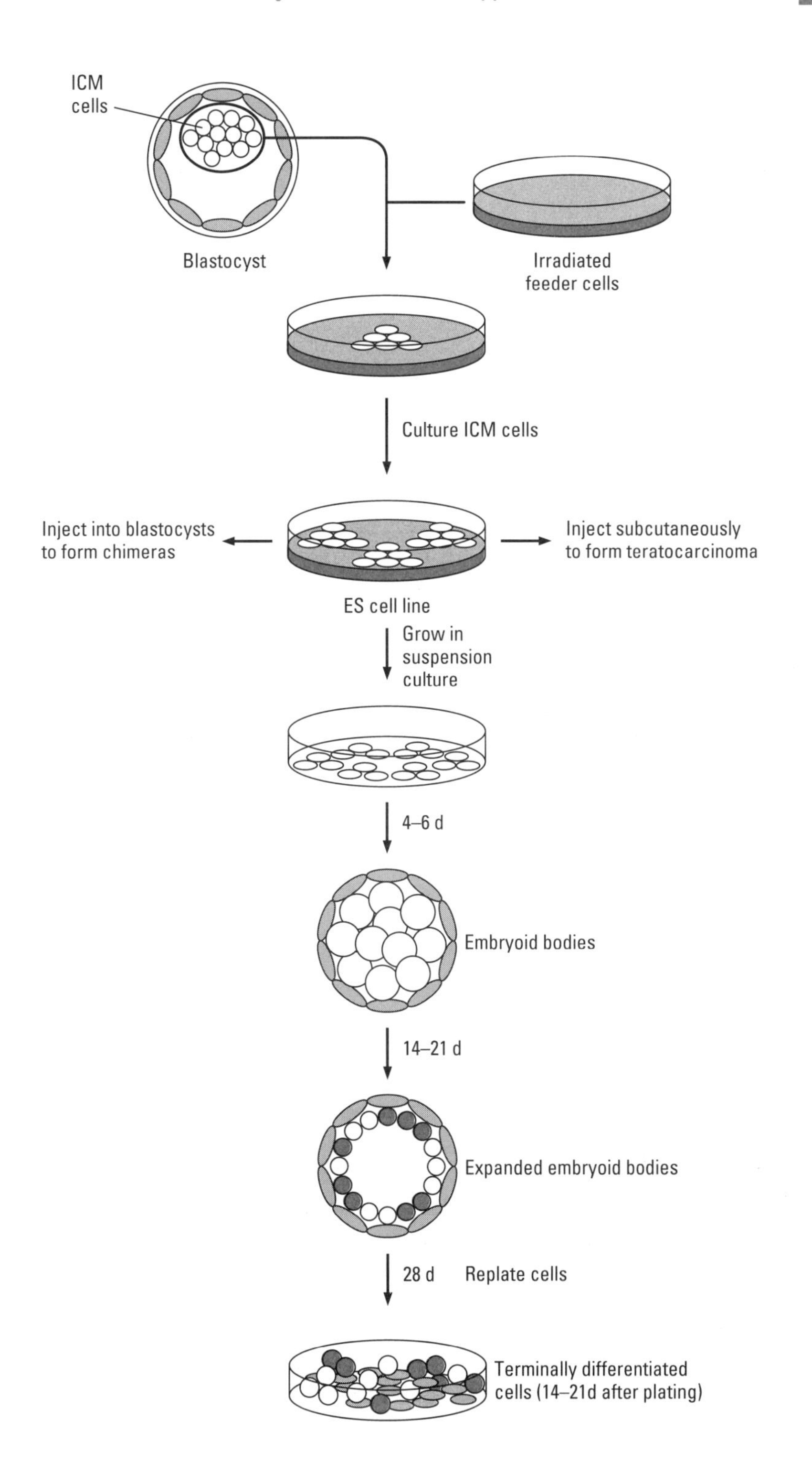
ICM
cells
Blastocyst
Irradiated
feeder cells
Culture ICM cells
Inject into blastocysts
to form chimeras
Inject subcutaneously
to form teratocarcinoma
ES cell line
Grow in
suspension
culture
4–6 d
Embryoid bodies
14–21 d
Expanded embryoid bodies
28 d
Replate cells
Terminally differentiated
cells (14–21d after plating)

into a blastocyst to produce a chimera (Figs. 6.11 and 6.12). Most and in many cases all, tissues within these chimeric embryos contain ES-derived cells, although in any given chimera the proportion of ES-derived cells varies. Thus, when ES cell chimeras are bred, many are found to have

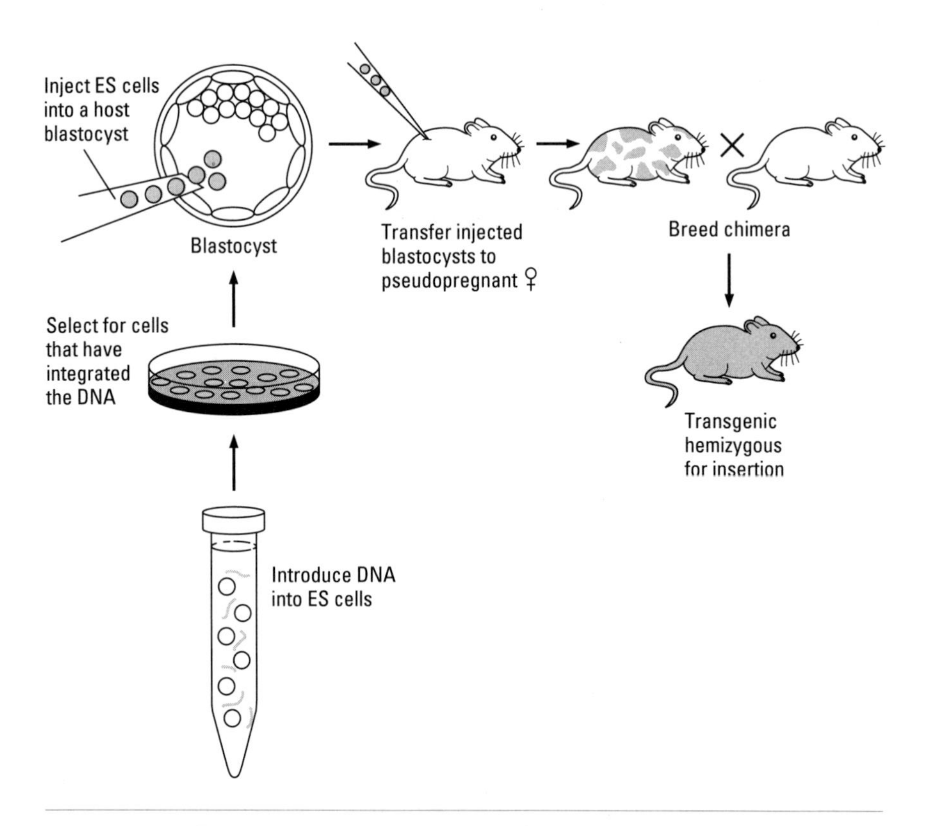

FIGURE 6.10 Procedure for making transgenic mice using ES cells. DNA is typically introduced into cells by electroporation (see Chapter 7). Rapidly growing ES cells are treated to make a single cell suspension, mixed with DNA (~40 μg/ml) and given an electric pulse to introduce the DNA into the nucleus. As described in Chapter 5, the DNA vector usually contains a selectable gene such as *neo*. The electroporated cells are therefore grown in G418 to select for cells that have integrated and are expressing *neo*. $G418^R$ cells from one clone are then used to make chimeras as described in the legend for Figure 6.1. Blastocyst injection is shown. The chimeric animals that are born are then bred, as described in the legend for Figure 6.8, to produce ES cell-derived offspring. About 50 percent of these mice should be hemizygous for the new inserted DNA.

ES-derived germ cells. The genotype of the host blastocyst, and perhaps of the ES cell line, influences the percentage of chimeras that contain an ES cell contribution to the germline. ES cells grown in culture can give rise to a mouse by first producing ES cell chimeras and then breeding the

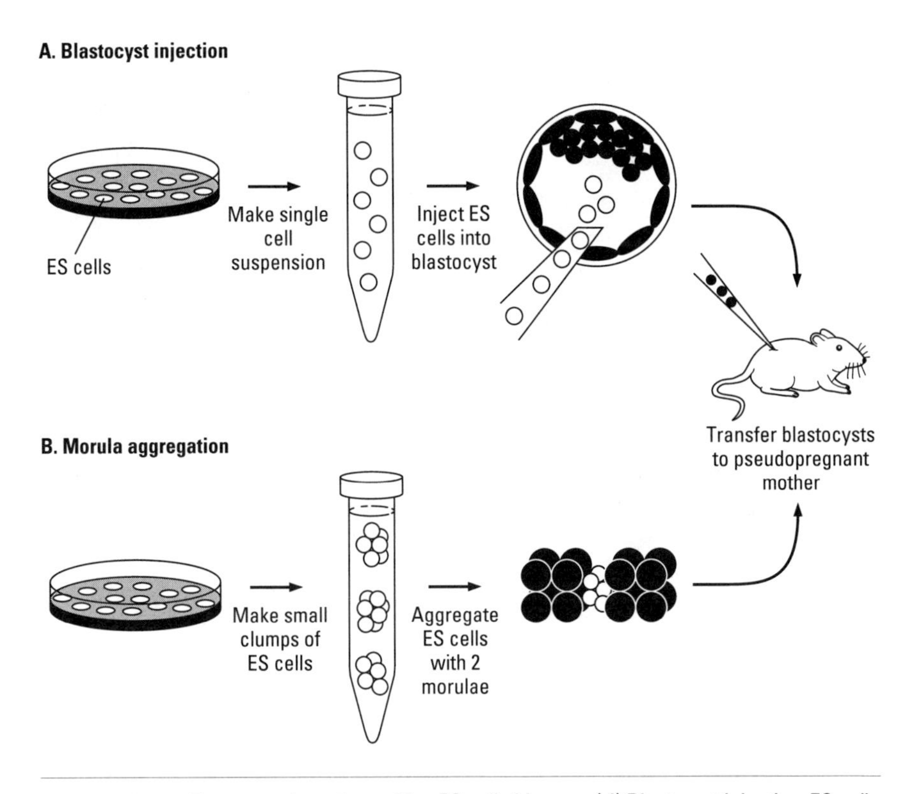

FIGURE 6.11 Two procedures for making ES cell chimeras. (*A*) Blastocyst injection: ES cells growing in an undifferentiated state are treated to give a single cell suspension. Blastocysts are isolated as described in the legend for Figure 6.7. The blastocysts of one genotype and ES cells of a second genotype are placed under a microscope with two micromanipulator setups. One micromanipulator is used to hold the blastocyst with a holding pipette. The other micromanipulator is used to suck up ES cells and inject 10 to 15 of them into the blastocoel cavity of the blastocyst with a finely drawn-out injection pipette. The blastocysts are then transferred to the uterus of a 2.5-day pseudopregnant mother. (*B*) Morula aggregation: ES cells growing in an undifferentiated state are treated to produce small clumps of approximately 10 to 15 cells. Small depressions are made in a dish and into each depression is placed a clump of ES cells with a morula on each side of it. The morulae are prepared as described in the legend for Figure 6.7. This sandwich of ES cells between two morulae forms a single aggregate and develops into a large blastocyst during the next day in culture. The blastocysts are then transferred to a 2.5-day pseudopregnant mother. In both methods approximately 50 percent of the embryos should reach term and approximately 50 percent of these should be chimeric. The ES cell contribution to a mouse can be as high as 80 percent of the cells.

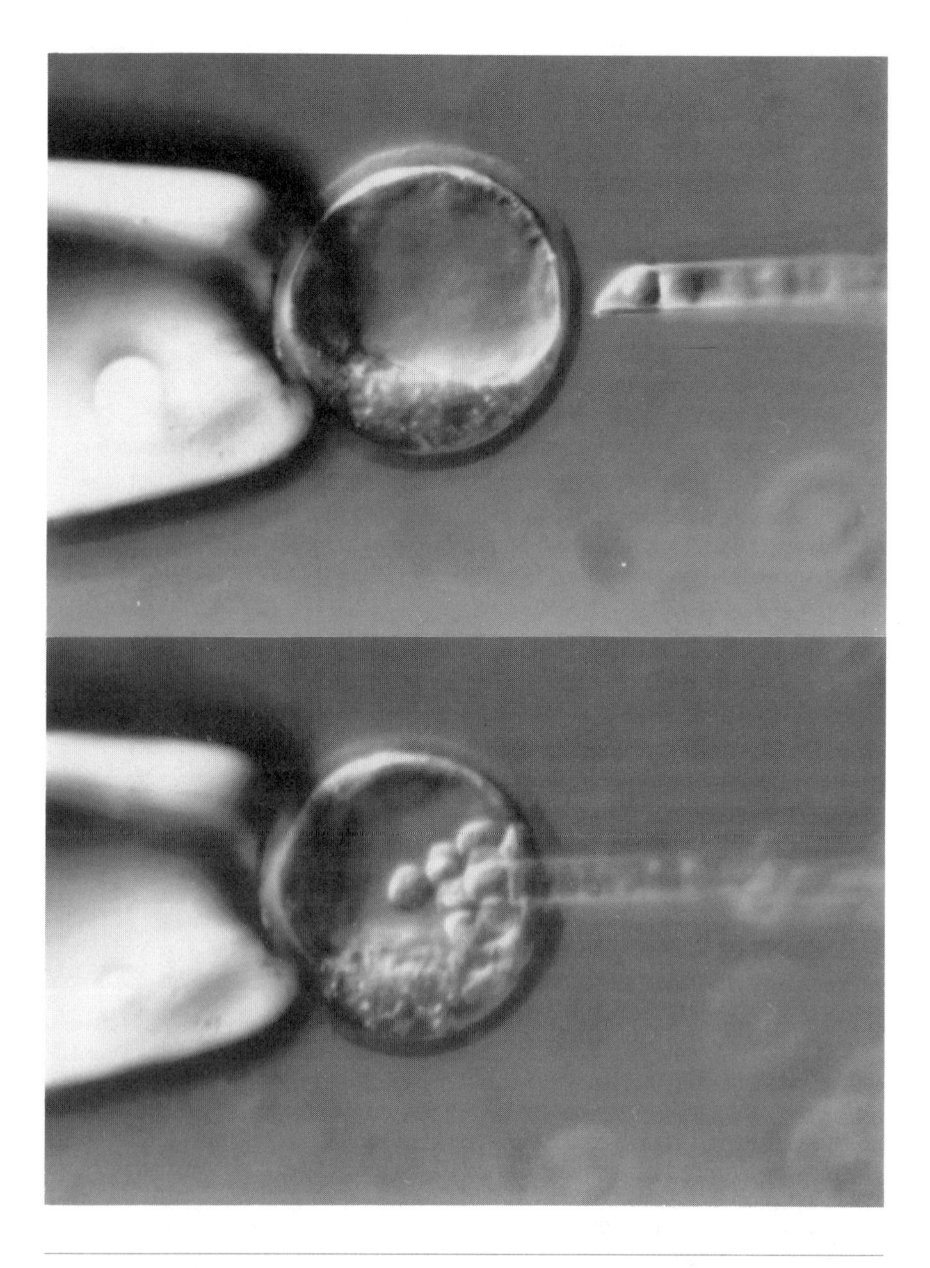

FIGURE 6.12 Microinjection of ES cells into a mouse blastocyst. *Top*: The holding pipette (left) is keeping the blastocyst (middle) in place while the injection pipette containing ES cells (right) is brought towards the embryo. *Bottom*: The injection needle is inside the blastocyst and ES cells have been released into the blastocoel cavity. The ICM can be seen on the inside of the blastocyst at the bottom. (*Photographs courtesy of Anna Auerbach.*)

chimeric animals to produce offspring totally derived from ES cells. Genetic markers such as coat color can be used to distinguish ES-derived mice, as described for embryo chimeras in the previous section.

One practical aside is that for most experiments ES cell lines derived from male embryos are used. One reason for this is that the karyotype of male cell lines is more stable than female lines. This seems to be because the two X chromosomes are unstable and one X chromosome tends to be lost, whereas the XY complement tends to be stable. The second reason for using male cell lines is that they produce a higher number of germline chimeras, or chimeras with an ES cell contribution to the germline. This is because when male ES cells are injected into a female blastocyst, the ES cells can convert the embryo to a male if the ES cell contribution to the somatic cells of the gonad is high. Furthermore, the female cells cannot form functional male germ cells, and thus the entire germline is derived from the ES cells. One indication that a male ES cell line can give chimeras with an extensive ES cell contribution is that the ratio of male to female chimeras born is skewed toward males. Finally, it is more efficient to breed males since they can be bred with numerous females during the period of time a female requires to produce a single litter.

ES cells can be induced to differentiate either as tumors in mice or in culture by changing the growth conditions. When injected subcutaneously into mice, ES cells form tumors, called teratocarcinomas, that contain a wide variety of differentiated cell types. In fact, the proliferating stem cells of spontaneous teratocarcinomas in mice can be established in culture as cell lines, called embryonal carcinoma (EC) cell lines, that have characteristics similar to ES cells. Although EC cell lines rarely maintain a normal karyotype, some EC lines can contribute extensively in chimeras.

In culture, ES and some EC cells can be induced to develop through a set of differentiation events that is very similar to the early postimplantation embryo (see Fig. 6.9). ES cells therefore can be used for certain experiments as an in vitro model system of early postimplantation embryogenesis. To promote this type of differentiation, ES cells are placed at a high density in the absence of LIF or feeder cells in culture dishes to which they cannot adhere. The cells growing in suspension then clump together and form large aggregates called **embryoid bodies.** Within a few days the cells on the outside of the clumps differentiate into primitive endoderm similar to the outer layer of the ICM of a 4.5-day embryo. A cavity then forms in the embryoid bodies and they begin to resemble an egg cylinder stage embryo. With time, mesodermal cells delaminate from the ectoderm, but normal gastrulation does not occur. Further differentiation of the ectoderm and mesoderm then can be induced by plating the cells on a tissue culture dish. After a few weeks many differentiated cell types, such as muscle, nerves, and skin, can be recognized in these cultures. This in vitro differentiation system offers some advantages over studying embryos. Homozygous mutations can therefore be made in ES cells by targeting both alleles of a gene and then determining the phenotypic effects of the mutation on development in vitro.

Note

1. Evans, M. J., and M. H. Kaufman (1981). Establishment in culture of pluripotential cells from mouse embryos. *Nature* 292: 154–156. Martin, G. R. (1981). Establishment of pluripotential cell lines from embryos cultured in medium conditioned by teratocarcinoma stem cells. *Proc. Natl. Acad. Sci. USA* 78: 7634–7638.

Suggested Readings

Bradley, A., Evans, M., Kaufman, M. H., and Robertson, E. (1984). Formation of germ-line chimeras from embryo derived teratocarcinoma cells. *Nature* 309: 225–256.

Bradley, A. (1990). Embryonic stem cells: Proliferation and differentiation. *Curr. Opin. Cell Biol.* 2: 1013–1017.

Doetschman, T. C., Eistetter, H., Katz, M., Schmidt, W., and Kemler, R. (1985). The *in vitro* development of blastocyst-derived embryonic stem cell lines: Formation of visceral yolk sac, blood islands and myocardium. *J. Embryol. Exp. Morph.* 87: 27–45.

Gossler, A., Doetschman, T., Korn, R., Serfling, E., and Kemler, R. (1986). Transgenesis by means of blastocyst-derived embryonic stem cell lines. *Proc. Natl. Acad. Sci. USA* 83: 9065–9069.

Hogan, B., Costantini, F., and Lacy, E. (1986). *Manipulating the mouse embryo: A laboratory manual.* Cold Spring Harbor, N.Y.: Cold Spring Harbor Laboratory.

Jaenisch, R. (1988). Transgenic animals. *Science* 240: 1468–1473.

Robertson, E. J. (ed.) (1987). *Teratocarcinomas and embryonic stem cells: A practical approach.* Oxford: IRL Press.

Rossant, J., and Pedersen, R. A. (eds.) (1986). Experimental Approaches to Embryonic Mammalian Development. Cambridge University Press.

7

GERMLINE GENE TARGETING IN MICE

The importance of making mutant organisms was introduced in Chapter 5, and the principles for making mice with defined mutations was described in Chapter 6. These two themes will be further developed in this chapter, using examples from actual research on gene function. The importance to medicine of germline gene targeting of mice and other future challenges of gene targeting are discussed in the last section of this chapter.

Three Approaches to Studying the Function of a Gene

For an experimental scientist, there are several levels at which the function of a gene must be explored to elucidate its contribution to the normal development and functioning of an organism. These levels are (1) descriptive studies of what cell types produce and/or interact with the gene product, (2) in vitro biochemical analyses of the properties of the gene and its protein product, and (3) studies of the biological function of the gene in living cells and ultimately in the whole organism.

At the first level, it is essential to determine when and where during development, as well as in the adult organism, a gene is expressed. This information provides clues to the processes in which the gene might be involved and, most important, which cell types should be further examined. At the second level, the biochemical properties of a gene must be understood. For example, a biochemical analysis of how the expression of a gene is regulated will determine which factors can stimulate or inhibit its transcription in a test tube. A biochemical analysis will also include studies of the purified protein product of a gene to determine its functional properties, for example, to define domains where the protein interacts with other proteins or DNA. Such in vitro studies are useful since they can often be technologically more feasible than studying the role of a gene in a living organism. The simplicity of in vitro studies stems from the fact that many, and in some cases all, factors involved in the reactions can be precisely defined and manipulated. The results of a biochemical analysis can thus reveal what a protein is capable of performing under the conditions tested; this can often lead to important clues to the possible biological function of a gene. Biochemical analyses are limited, however, in trying to replicate in vitro all of the situations a molecule normally encounters in a cell or organism. Thus such studies often underestimate the full repertoire of functions a protein can perform.

The most complex level of research involves studies in vivo. Because of the complexity, it is often useful to first study the function of a gene in a single cell type in culture before tackling the role of the gene in the whole organism. In contrast to in vitro systems, the multitude of molecules that make up a cell are not defined, and this often leads to some discrepancies between in vitro and in vivo results. One reason for the different results may be that in a test tube a protein is often asked to perform a function on its own, whereas in a cell it may have many helpers that assist in carrying out the function. If this is the case, then if a domain of a protein is mutated and the function of the protein tested in vivo, the mutant protein may perform almost as well as if the domain were normal because other factors are making up for its absence (Fig. 7.1*A,C*). In vitro, on the other hand, the mutant protein may not be able to function at all (Fig. 7.1*B,D*). Discrepancies also arise between descriptive data on where a gene is expressed and

FIGURE 7.1 (***Right***) Complexity of protein–protein interactions involved in transcription in vitro and in vivo. (*A*) Normal in vivo situation in cells. A stretch of DNA (double helix) is shown bound by two proteins (A and C) that are in turn bound by a third protein (B). Protein B acts as a bridge for proteins A and C, connecting them with the transcription machinery (RNA polymerase and other accessory proteins). A high level of transcription is stimulated when A, B, and C are in contact with gene *X* as shown. (*B*) Test system in vitro that includes only proteins B and C. Proteins B and C can stimulate a moderate level of transcription in vitro when bound to gene *X*. (*C*) A mutant form of protein C has been introduced into cells by gene targeting. The region of protein C that binds protein B has been mutated so that it can no longer bind protein B. Nevertheless, protein A can bind gene *X* and stimulate transcription through binding protein B. (*D*) The mutant form of protein C is tested *in vitro*. Protein C can bind to gene *X* but cannot stimulate transcription since it cannot bind protein B.

in vivo mutant phenotype data on which tissues require the gene product. In the majority of cases to date, the tissues showing defects in mutants are only a subset of the tissues that express the genes. One explanation for this discrepancy is that since most genes are members of gene families made up of similar genes that perform similar functions, in the absence of

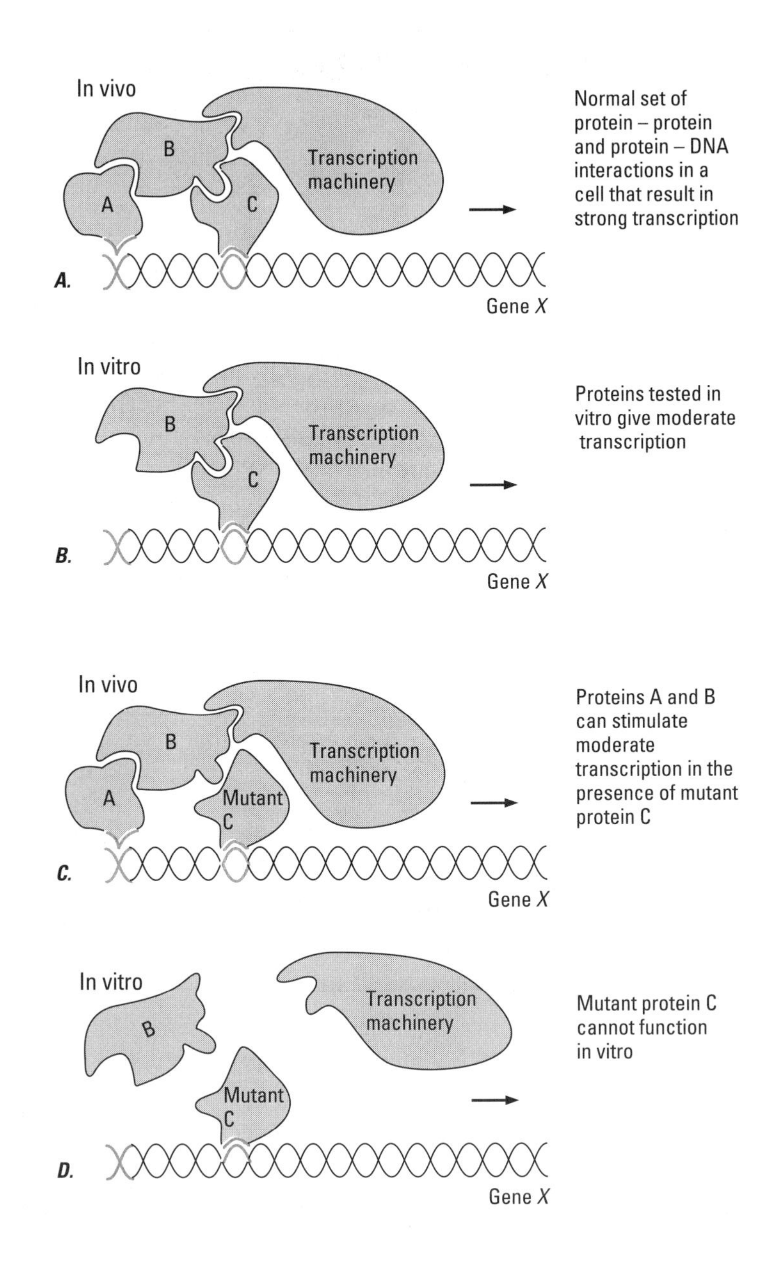

one gene the others can compensate for its loss. This idea of "backup systems" in vivo is a recurrent theme. It is essential to have descriptive, biochemical, and genetic data, since the results of all these types of studies are complementary and can lead to further investigations at all levels.

Why Gene Targeting and Not Random Integration?

One of the key reasons gene targeting is so important is that it can simplify studies done in cells or in animals because the investigator can predetermine exactly what genetic change is made. This is in contrast to experiments in which DNA is randomly inserted into the genome, where many factors cannot be strictly controlled. To further explore the question of why germline gene targeting is important, it is useful to go back to transgenic mice made by zygote injection and think about the kinds of questions that can be tackled best using this approach in comparison to germline gene targeting. It should then become clear why germline gene targeting in mice was a major technological breakthrough.

Zygote injection is currently limited to introducing a fragment of DNA into the genome by nonhomologous recombination. Information can be added, but it cannot be removed or substituted. Thus, one is limited to asking, What can the extra piece of DNA do? In one type of experiment the function of the piece of DNA itself can be tested, for example to examine whether it has the necessary recognition sequences to initiate transcription in particular cell types (Fig. 7.2). In another type of experiment, the function of the gene product can be tested (Fig. 7.3). An example of this would be making transgenic mice, and testing them for whether a certain protein can promote the formation of tumors by expressing it in some or all cells and observing whether tumors form at a higher frequency in the transgenics than in wild type mice. In a third type of experiment, the function of a particular cell type can be addressed. An example of this would be to make a transgenic animal expressing a toxic protein in specific cell types to determine the effect on the animal of ablating the cells.

Although on paper these experiments sound straightforward and easy to interpret, in practice they are not. For example, where does one get a DNA fragment with a promoter that will express in a specific tissue of interest, at the time of development of interest, and at the level required? Even when such DNA fragments can be obtained, a major problem is the random nature of the integration events. Some regions of the genome can act to inhibit transcription, whereas others can promote transcription.

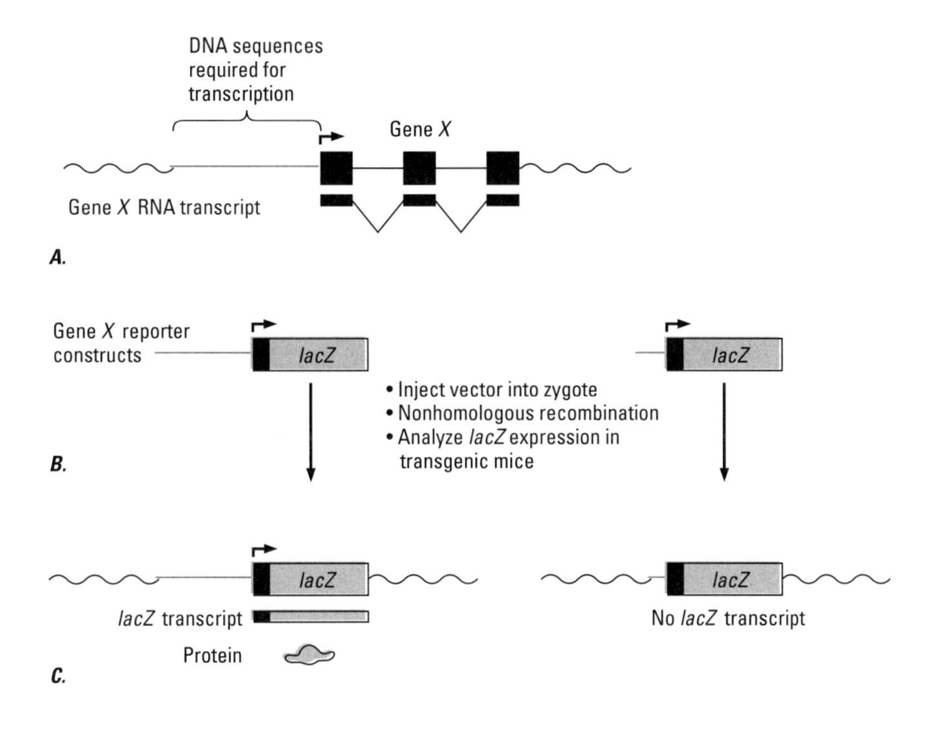

FIGURE 7.2 Use of transgenic mice to identify DNA sequences capable of directing transcription to particular tissues. (*A*) Gene *X*, consisting of three exons (filled squares) and an upstream regulatory region (green). In the RNA transcript from gene *X*, the V's between exons indicate where splicing takes place. (*B*) Two "reporter" constructs that include the bacterial gene *lacZ* as the reporter gene. This gene encodes the β galactosidase protein. Cells expressing this protein can be detected by histochemical staining (see Fig. 7.11). The two vectors also include either the complete regulatory region (left) or a part of it (right) and a portion of the first exon. Each vector is injected into zygotes, and the vectors integrate into the genome by nonhomologous recombination. Transgenic animals that are born are detected by the presence of lacZ sequences in their DNA. (*C*) Each vector is integrated into the genome. Only the transgene that includes the complete DNA regulatory region expresses lacZ transcripts and β galactosidase protein.

These effects are called **position effects.** This limitation requires that each founder animal be treated as independent and that information from at least three founder animals be obtained and compared to determine what characteristics are common to all. Another problem that can arise from random integration is that the new DNA can inactivate an endogenous gene, causing a new mutation. In fact, disruption of an essential gene occurs in approximately 10 percent of transgenic mice made by zygote injection.

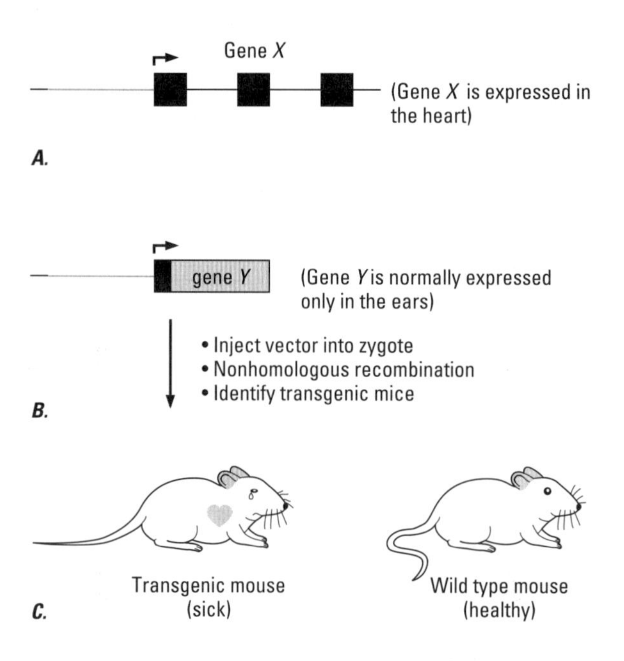

FIGURE 7.3 Use of transgenic animals to determine the biological consequences of expressing a gene inappropriately. (*A*) Gene *X* expressed in the heart shown as in Figure 7.2. (*B*) Construct in which the DNA regulatory region (green) of gene *X* is linked to the protein coding sequences of gene *Y*. Gene *Y* is normally expressed only in the ears. The construct is injected into zygotes to obtain transgenic animals. (*C*) If gene *Y* (green) is expressed in the heart as well as in the ears of a transgenic mouse, the animal becomes ill.

Gene targeting can complement and extend these studies. The most important feature of gene targeting is that it can remove and/or substitute for information already present. Thus, in the case of identifying DNA regulatory elements, it would be informative to remove an element that from other studies appears to be essential and test whether it is actually required for transcription at its normal chromosomal location. It may be that there are many other similar elements in vivo that collaborate with and augment the activity of that particular element. Similarly, a good way to faithfully replicate the expression pattern of a given gene is to insert other coding sequences into the endogenous gene by gene targeting.

The most important use of gene targeting for experimental studies is making change-of-function mutations that alter the biochemical properties of a gene product. These altered properties can range from no activity (a **null** mutation) to partial activity (a **leaky** mutation) to a new or stronger activity (**dominant** mutation). Some types of mutants can be made by zygote injection. The previously discussed cases of expressing genes in new tissues represent dominant mutations. Null mutations using zygote injection also have been attempted, using vectors that express gene products that can inhibit the normal function of an endogenous gene. These are called **dominant/negative** mutations. One example of this is the "antisense" approach, in which a vector is made that will transcribe the antisense, or opposite, strand of the coding DNA to the normal gene. The antisense RNA that is produced forms a double helix with the normal transcript, thus inhibiting its translation into protein (Fig. 7.4). A number of complicating factors in this approach make the results difficult to interpret. The major problem is expressing the vector in the same cells at the same time as the normal gene, as well as expressing it in sufficient excess of the normal transcript to drive the annealing reaction of the two RNA molecules to completion. It should be apparent by now that the most straightforward way to make a null mutation would be to use gene targeting to directly remove a gene from the genome.

More on the Importance of Mutant Organisms

Why are null mutations important? At the simplest level they provide a base line for the development or functioning of an organism in the absence of a particular gene. This will answer the question, Is the gene essential for life? If it is essential, then experiments can proceed to determine what process is interrupted in the absence of the gene. If the gene is not essential, then the question becomes, Why not? Does the gene not function the way that was predicted from the biochemical analysis and from its expression pattern, or are there other backup systems that can partially compensate for the loss of the gene? If the latter is the case, then the other genes must be identified and also removed to test the combined function of the genes.

Why are leaky mutants important? For genes that are essential during early embryogenesis, it is important also to make partial loss-of-function mutants to test for multiple functions of the gene. For example, a mutation could be designed that allows the animal to develop normally through the first critical process in which the gene functions to test for the next critical function later in development. Alternatively, leaky mutations can test the biological significance of particular protein domains.

Why are dominant mutations important? In the case of mutants in which a gene is expressed in new cell types, the question being addressed is the functional potential of the gene. For example, Can a gene cause a tumor, or can a gene change the differentiation pathway of a cell? If the

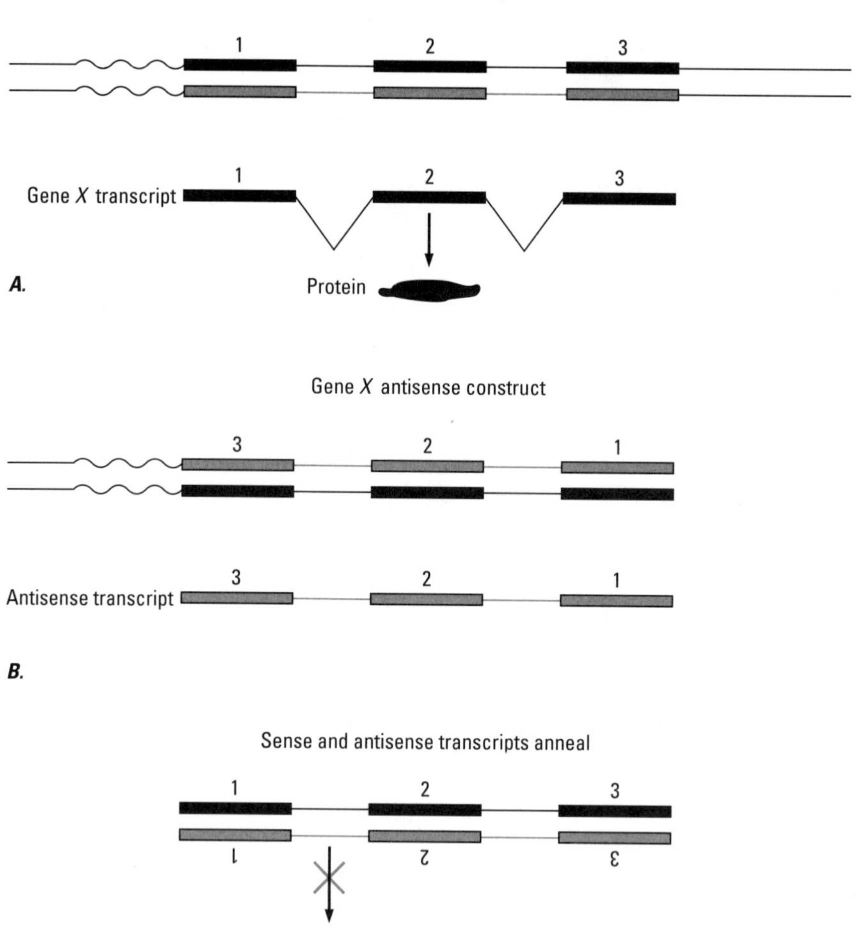

FIGURE 7.4 Expression of antisense RNA can interfere with the expression of a gene. (*A*) *Top:* The two strands of DNA of a three-exon (filled rectangles) gene *X*. The lower strand is used as the template for transcription. *Bottom:* Transcript and protein from gene *X*. (*B*) *Top:* Antisense construct for gene *X*. The DNA regulatory region of gene *X* (wavy lines) is joined to gene *X* sequences that have been reversed in their direction of transcription. Bottom: Transcript that would be produced from the antisense construct. Note that compared to the normal gene, the opposite strand of DNA is used as the template for transcription. (*C*) Annealed sense and antisense transcripts.

answer is yes, it suggests that the gene plays some sort of regulatory role in directing the development of cells. When this is the case, the gene often turns out to be a factor that regulates the activity of other genes. Dominant mutations that do not alter the expression pattern of a gene, but instead alter or augment the activity of the protein product by changing selected amino acids, are often designed to test the function of the particular domains that are altered (as in Fig. 7.1). This type of analysis of the functional properties of a protein is referred to as **structure/function** analysis and can be most effective if done in vitro or in cells in culture where there are fewer variables to contend with.

From the analysis of mutant phenotypes from the germline targeting experiments that have been reported to date, two themes are emerging that have important implications for these types of studies of in vivo gene function. One is that the tissues affected by null mutations represent only a subset of all the tissues that express the gene. This lends support to the idea of backup systems. The second theme is that genes play many different roles in different tissues at different times in development. This lends support to the idea that gene products do not perform their functions alone but act in concert with many other proteins, and since the expression of the other proteins often varies from cell type to cell type, the role of the gene in question also varies.

Because of this complexity of gene function, it is necessary to make animals with different kinds of mutations, and gene targeting offers the only method for making predetermined mutations in which the maximum number of potential complicating factors can be controlled. The targeting approaches that have been used extensively to date in ES cells all include insertion of a selectable gene into the gene of interest. This insertion type of mutation can be effective in making null mutations and dominant mutations involving altering the expression of a gene. However, insertion mutations can be problematic when it comes to making more subtle mutations such as single base pair changes to make leaky and dominant gain-of-activity mutations. For these types of mutations, pop-in/pop-out or direct gene replacement procedures will have to be perfected.

General Considerations

As was discussed in Chapter 6, the present method of choice for germline gene targeting is making the directed gene mutation in ES cells and then transmitting the mutation into mice by making germline ES cell chimeras.

The mutation is made in ES cells by introducing a gene targeting vector into ES cells and allowing the cells to undergo a homologous recombination event that replaces the normal endogenous gene with the introduced mutant copy. As described in Chapter 5, the technological breakthroughs that have made this feasible in mammalian cells have been (1) enrichment schemes that select preferentially for the targeted cell clones and (2) screening procedures using PCR that allow thousands of colonies to be screened for rare homologous recombination events.

The special techniques that have been developed for gene targeting in mammalian cells, described in Chapter 5, apply to ES cells. In general, replacement rather than insertion vectors are used (Fig. 7.5). From a practical point of view, ES cells differ from most somatic cells in two important ways. The first is that they are much smaller. This is an advantage, since from a tissue culture dish tenfold more cells can be recovered. The second difference, a disadvantage, is that ES cells have very specific growth requirements (described in Chapter 6), such as a feeder layer or LIF and a rigid protocol of passing the cells frequently. Careful culturing of the cells is critical for the targeted cells to maintain their ability to contribute to the germline in ES cell chimeras.

The first factor to consider when designing a homologous recombination experiment is whether the gene to be targeted is expressed in ES cells. This dictates the type of targeting vector that should be constructed. If the gene is expressed, then the most efficient type of vector is probably one designed on the principle of a promoterless selectable marker gene (see Fig. 7.6), since these types of vectors in general give the greatest level of enrichment. The bacterial *neo* gene is a good choice for the selectable marker, since G418 selection is simple and effective in ES cells and does not effect the totipotency of the cells. If the ES cells are grown on a feeder layer rather than in LIF, then $G418^R$ feeder cells must be used. $G418^R$ STO cell lines can be obtained by isolating $G418^R$ cell clones transfected with a vector that expresses the *neo* gene. Alternatively, $G418^R$ primary fibroblasts can be isolated from transgenic mice that express *neo*.

If the gene to be targeted is not expressed in ES cells, then a vector can be designed on the principle of positive–negative selection (PNS; see Fig. 7.7). One factor that may be critical for targeting genes that are not expressed in ES cells is the choice of promoter for the selectable genes. As discussed earlier in this chapter, not all DNA is equal, and some regions of the genome are transcriptionally "silent." Thus, it is possible that some nonexpressed genes will actively inhibit expression of a newly introduced promoter intended to express *neo*. If this is the case, then any cells that have undergone homologous recombination of the vector will not be able to express the *neo* gene and thus will be killed by the G418 selection. In an attempt to solve this problem, a number of *neo* vectors containing different

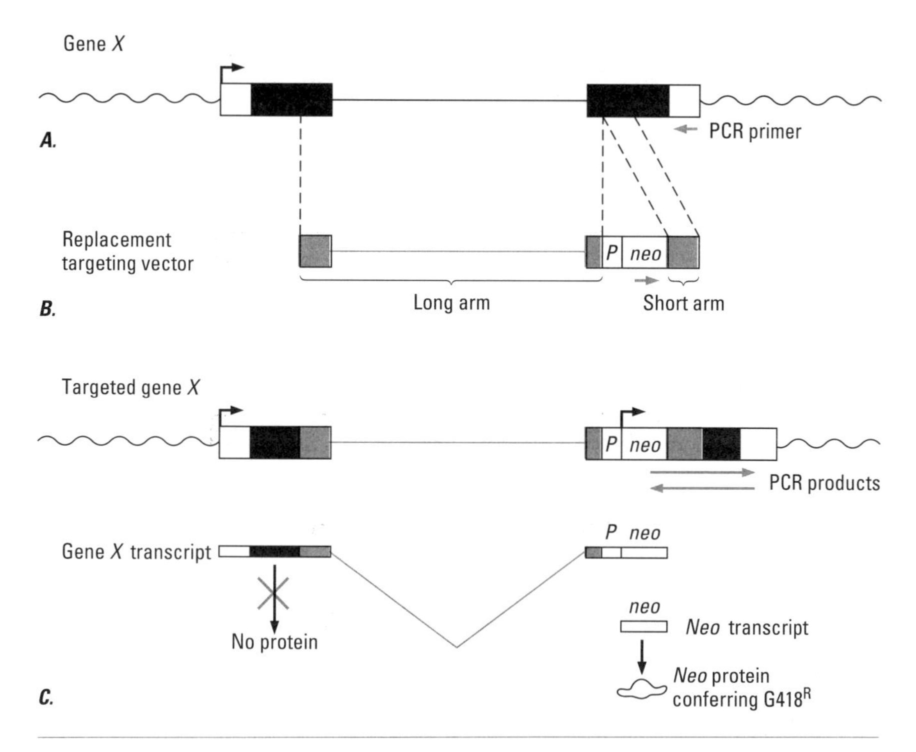

FIGURE 7.5 Gene targeting in ES cells using a basic replacement vector. (*A*) Two-exon gene *X*. Exons are indicated by rectangles with filled-in areas indicating protein coding sequences. The primers used for PCR in screening for homologous recombinants are shown as short green arrows. The start and direction of transcription is indicated by an arrow. (*B*) Replacement targeting vector with long and short fragments of gene *X* (long arm and short arm, respectively) surrounding the *neo* gene driven by a transcription promoter (P). (*C*) Structure of gene *X* following homologous recombination and replacement of gene *X* sequences by the targeting vector. The PCR products would be made using the two primers indicated (one that anneals to *neo* sequences and one that anneals to gene *X* sequences not present in the targeting vector). From the transcripts that would be produced from the promoters of the genes *X* and *neo,* no functional protein *X* would be made, whereas the *neo* protein product would be made and confer G418 resistance to the cells.

promoters have been constructed with the idea that a really strong integration-site-independent promoter will not be completely shut down even in silent genes. It is not yet clear whether any of the available promoters can in fact be expressed in all genes.

A second possible limiting factor in all targeting strategies is that some DNA may undergo recombination at very low frequencies. From the data to date, the frequencies of successful targeting events range at least three orders of magnitude. These numbers, of course, do not include genes for

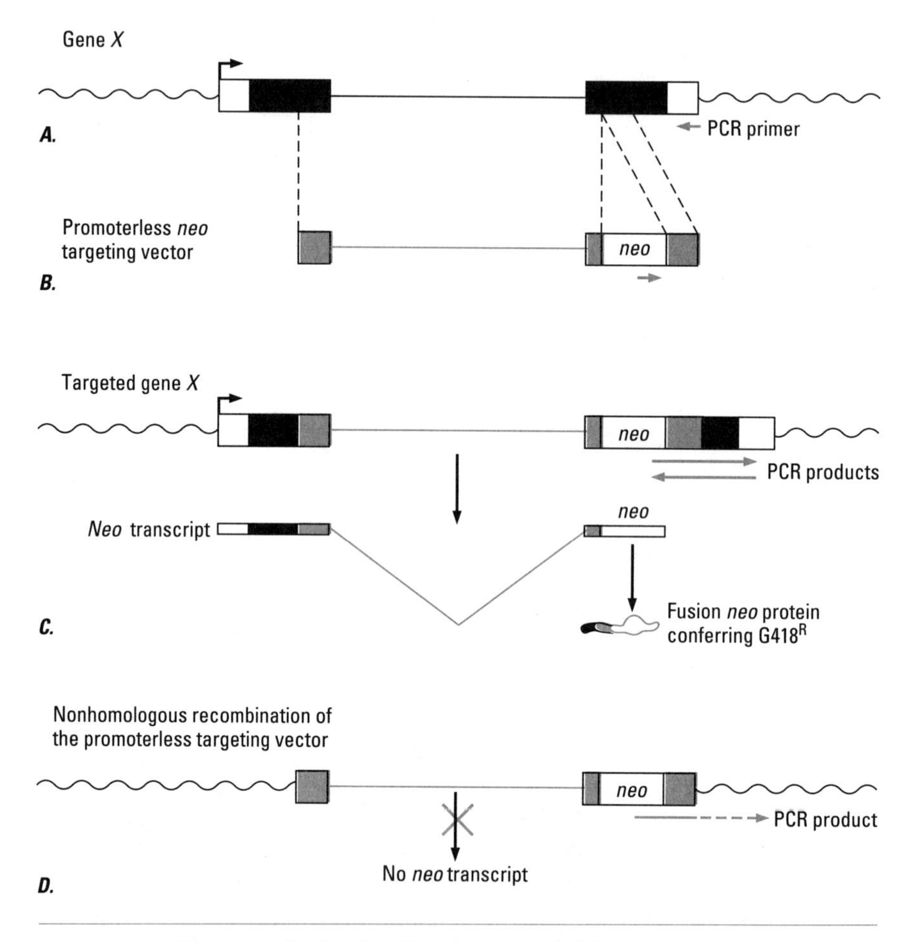

FIGURE 7.6 Gene targeting in ES cells using a promoterless *neo* targeting vector that enriches for cells that have undergone homologous recombination. (*A*) Gene *X* shown as in Figure 7.5. (*B*) Gene targeting vector similar to that in Figure 7.5*B*. The difference in this vector is that *neo* does not contain a promoter and the *neo* protein is made as a functional fusion protein with a portion of protein X. (*C*) *Top:* Structure of gene *X* following homologous recombination with the promoterless *neo* targeting vector. *Bottom:* Transcript and protein made from the targeted gene *X*. (*D*) Structure of the promoterless *neo* targeting vector following nonhomologous recombination with random genomic sequences. Unless the vector integrated by chance beside a promoter, no *neo* transcript would be made.

which it has not been possible to target. Thus the lower limit of the frequency of homologous recombination is not known. The only obvious possible solution to this problem is to construct targeting vectors containing different regions of a gene, in the hope that some fragments of DNA will be more recombinogenic than others.

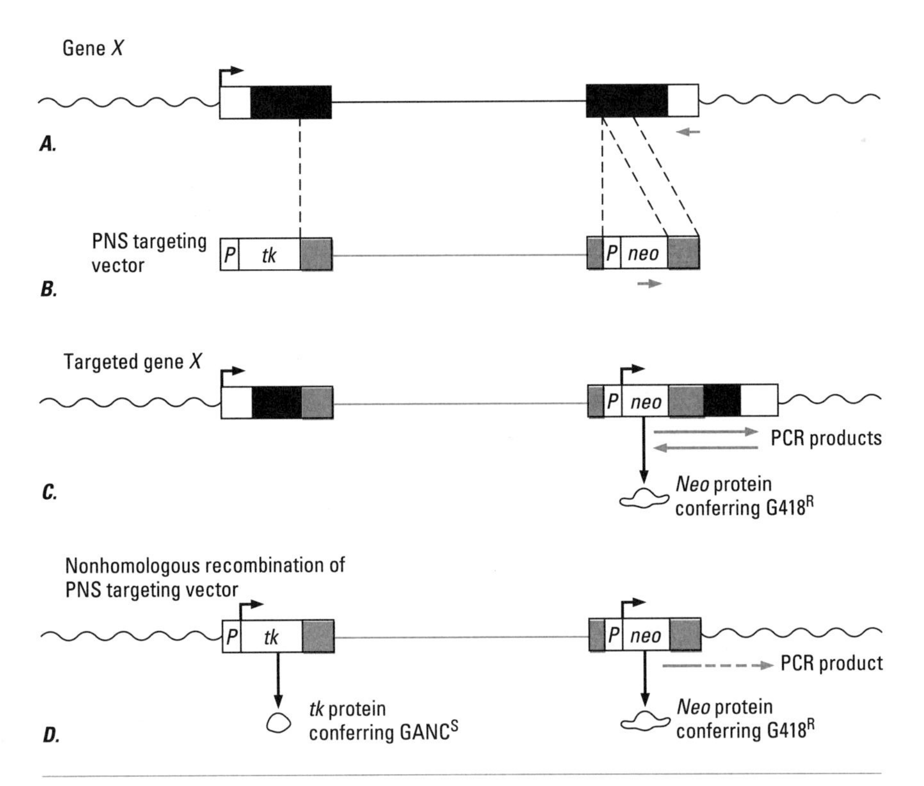

FIGURE 7.7 Gene targeting in ES cells using a targeting vector based on the positive–negative selection (PNS) enrichment scheme. (*A*) Gene *X* shown as in Figure 7.5. (*B*) Gene targeting vector similar to that in Figure 7.5*B*. The difference in this vector is that the *Herpes simplex* virus *tk* gene with a promoter is included at the end of one of the gene *X* homologous fragments. (*C*) Structure of gene *X* following homologous recombination with the PNS targeting vector. The structure is the same as that shown in Figure 7.5*C*, since the recombination occurs within the gene *X* sequences and thus the *tk* sequences are lost. (*D*) Structure of the PNS targeting vector following nonhomologous recombination. Functional *neo* and *tk* genes will be integrated into the genome if the vector DNA is not broken. The *neo* protein confers $G418^R$ to the cells, whereas the *tk* protein confers sensitivity to GANC. Cells containing such a vector will be killed if they are grown in G418 and GANC.

A typical gene targeting experiment can be divided into a number of steps. The first is to design and make a targeting vector taking into consideration the enrichment schemes and possible limitations described above. The second step involves introducing the targeting vector into ES cells and selecting a population of cells that have taken up the vector (Fig. 7.8). A sample of ES cells that had been frozen soon after the cell line was established is thawed and grown up. The targeting vector is then introduced into the cells using electroporation. The targeting vector must be prepared by linearizing the

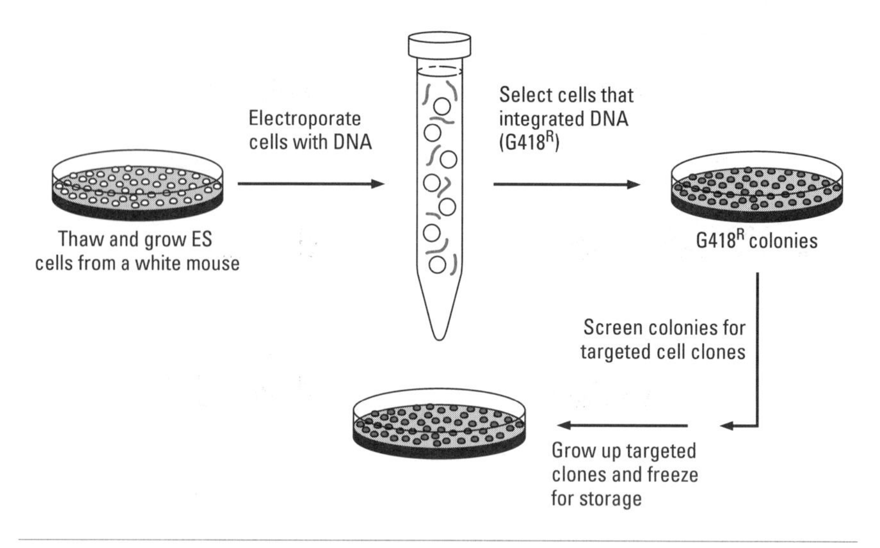

FIGURE 7.8 Scheme for electroporating ES cells and isolating cells containing a targeted mutation.

DNA, and the cells must be treated to give a single cell suspension. The DNA and cells are then mixed, put into a cuvette with electrodes on either side, and given a short pulse of an electric field. The DNA is driven across the cell membranes and taken up into the cell nuclei by approximately 10 percent of the cells. The cells are then plated on dishes either in LIF or on feeder cells that are G418^R. Two days later the selective medium, G418 or G418 and GANC, is put onto the cells. Within ten days colonies appear on the dishes. The normal ES cell plating efficiency, or number of cells that will grow after being replated, is 10 to 20 percent; and 50 percent of the cells are usually killed by the electroporation procedure. Furthermore, only approximately 0.1 percent of the electroporated cells stably integrate and express a marker gene. Thus, in an experiment starting with 5×10^7 cells and a standard *neo* vector, approximately 5000 G418^R colonies would grow. Using a vector with a promoterless *neo* gene, approximately 50 G418^R colonies would probably grow, and using the PNS strategy 50 to 500 G418^R/GANC^S colonies would grow. The percentage of targeted clones that would be obtained from such an experiment could vary from 0 to 50 percent, depending on the type of vector and particular endogenous gene sequences used.

The third step in a targeting experiment involves screening the drug-resistant colonies for clones that have integrated the vector by a homologous recombination event. Because the frequency of homologous recombination varies widely, and since it is not possible to predict a priori the targeting frequency of a given vector, it is helpful to design the vector so that PCR can be used to screen for the recombinants (Figs. 7.5, 7.6, 7.7). This requires that one of the genomic DNA fragments that flanks *neo* is less than 1 thousand base pairs (Fig. 7.5, short arm). One primer for the PCR reaction is then made to anneal to *neo* sequences and the other to endogenous gene sequences just outside the sequences included in the short arm of the vector.

To identify the targeted clones, a modified version of sib selection (see Chapter 5) can be used (Fig. 7.9). Instead of pooling whole *neo*^{R} or *neo*^{R} and GANC^S colonies, as in sib selection, the colonies are left intact, and only a part of each colony is picked using a pipette and suction under a microscope. The portions of up to 50 colonies are then pooled and the PCR reaction run on each pool. The remainder of the colonies are left to grow. When a positive pool is identified, portions of each single colony from that pool are then tested by PCR. A single targeted colony can be identified by this approach in less than one week. The targeted colony is then grown up and frozen in small samples to make a stock of cells for future use.

The advantages of this modification of sib selection for ES cells are twofold. First, not all ES colonies grow at the same rate and thus slower-growing colonies are not lost after pooling. The second and more important advantage is that since the ES colonies are left intact, they go through a minimum length of growth. This is critical since, as discussed in Chapter 6, the length of time in culture and number of subclonings is inversely correlated with the probability that a given ES cell clone will retain its ability to populate the germline in chimeras. Depending on the state of the starting population of ES cells used in an experiment, the percentage of cell lines that will populate the germline can vary from less than 10 percent to 100 percent.

The final step in germline targeting is to make ES cell chimeras that have an ES cell contribution to the germline (Fig. 7.10). Since not all subcloned cell lines will populate the germline, it is best to isolate 2 to 5 targeted cell lines and test them all in chimeras. Blastocyst injection has been the method of choice for obtaining chimeras because the percentage of animals that reach term is generally higher. Two ES cell lines derived from inbred mice have been used extensively for germline targeting experiments by many different laboratories. With these cell lines it has become apparent that the genotype of the host blastocyst has a strong effect on the percentage of chimeras that contain an ES cell contribution to the germline. Host blastocysts from outbred mice give the lowest numbers of germline chimeras, whereas blastocysts from particular inbred strains give high numbers. With some ES cell clones, most of the chimeras in an experiment can have an ES cell contribution to the germline. Since, for the reasons described in Chapter 6, male ES cell lines are normally

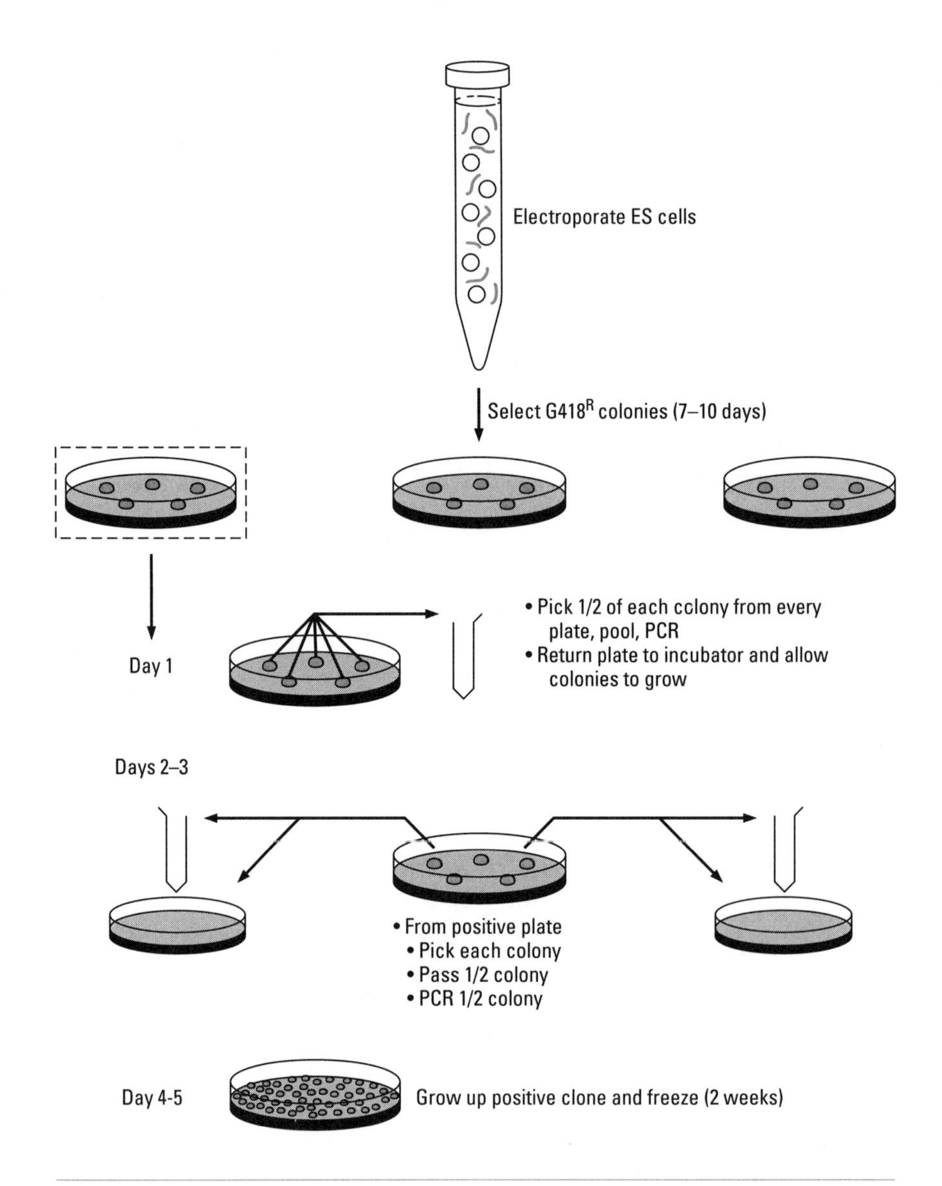

FIGURE 7.9 Strategy for screening ES cells by a modified version of sib selection, using PCR to identify targeted cell lines.

used, only the male chimeras are bred to produce mice heterozygous for the targeted mutation. Coat color markers are used as a convenient means of identifying ES cell derived mice and DNA isolated from tail biopsies is analyzed to identify the 50 percent of the animals that carry the targeted mutation.

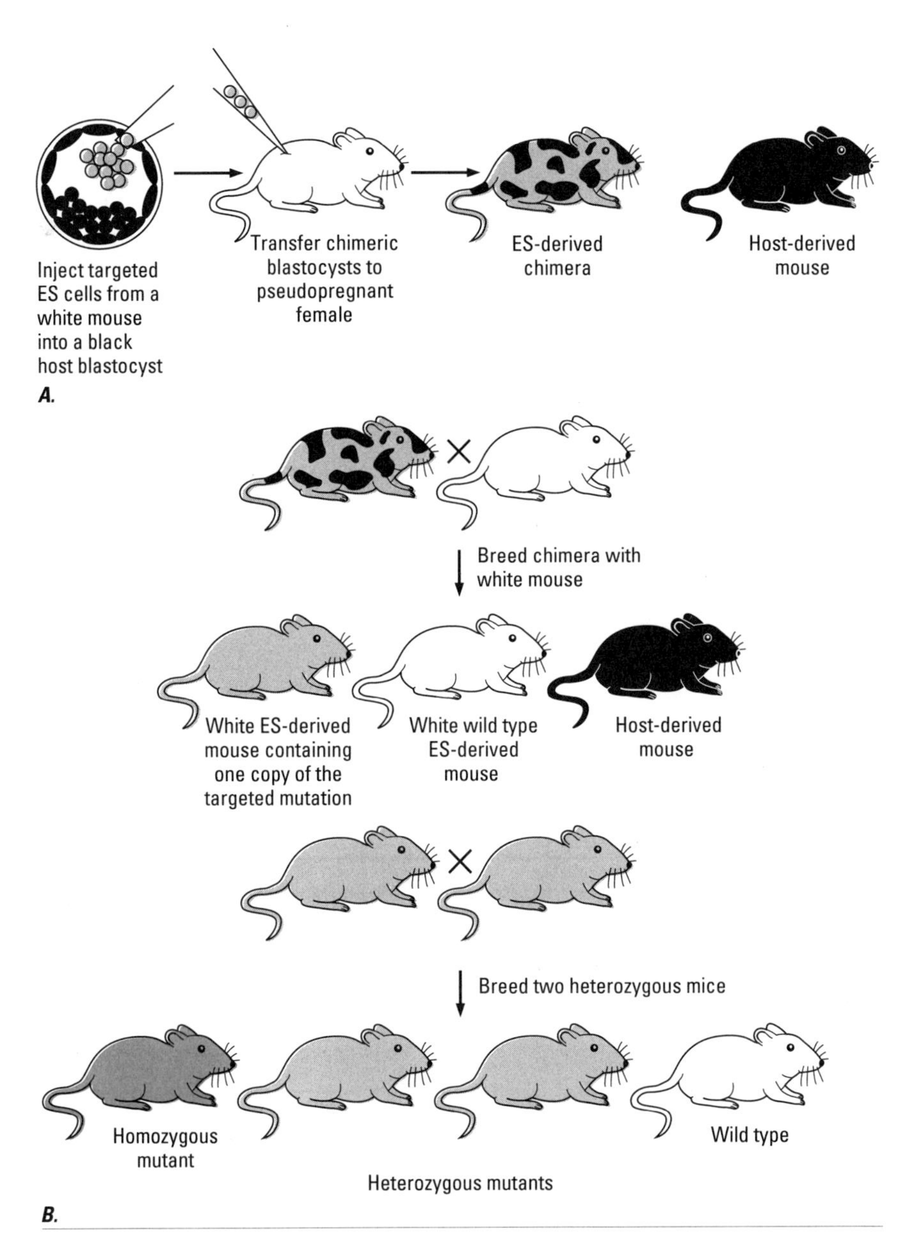

FIGURE 7.10 Strategy for obtaining mice homozygous for a targeted mutation from ES cells heterozygous for the mutation. (*A*) Production of ES cell derived chimeras, as described in the legend of Figure 6.11. The targeted ES cells and their derivatives in mice are shown in light green and the host blastocyst derived cells in black. (*B*) The male chimeras are bred to produce mice derived from the targeted ES cells (light green and white). The ES cell derived mice are analyzed to determine which contain one copy of the mutation (light green). The mice that are heterozygous for the mutation are bred, and one quarter of their offspring will be homozygous for the mutation (dark green).

A Specific Example

To illustrate the unpredictable nature of both the biological functioning of genes and the process of gene targeting, it is useful to examine a specific example. The mouse genes in question are called the *engrailed* genes, or *En-1* and *En-2* for short. These genes were isolated from mice based on their DNA sequence similarity to the *engrailed* (*en*) and *invected* (*inv*) genes of the fruit fly *Drosophila melanogaster*.

The *en* gene was identified in *Drosophila* in 1929 based on the phenotype of an *en* mutant fly that arose spontaneously.[1] The posterior half of the wings of flies carrying this *en* mutation develop abnormally whereas the anterior halfs develop normally. Based on the phenotypes of this and many other *en* mutants, a general hypothesis was put forward that the biological function of *en* is to direct, or select, the developmental pathway of groups of cells, called posterior compartments, that make up the posterior portion of every segment and of tissues, called imaginal discs, that give rise to the major parts of the adult body such as limbs, wings, and antennae.

In 1985 the *Drosophila en* gene and a gene, *inv*, with a similar DNA sequence were cloned.[2] The sequence of the protein coding regions of *en* and *inv* suggested that they code for transcription factors. Subsequent biochemical analysis on *en* protein confirmed this. Analysis of expression of *en* and *inv* in fruit fly embryos showed that they were expressed in all the cells of the posterior compartments, consistent with the *en* mutant phenotypes. The genes were also, however, found to be expressed in reiterated sets of neurons in the central nervous system. This suggests that there are other, as yet undiscovered, functions for *en* in the development of the adult nervous system.

The two mouse genes, *En-1* and *En-2*, were cloned based on their DNA sequence homology to *en* and were found to have protein coding regions that share a number of domains with *en* and *inv*.[3] The protein sequence conservation suggests that the mouse genes, like *Drosophila en*, act as transcription factors. The compelling question to be answered, therefore, was, Do the mouse genes also act to direct the development of posterior compartments?

The first indication that evolution had not strictly conserved the biological function of *en* in invertebrates and vertebrates came from the analysis of the expression of *En-1* and *En-2* in mice and other vertebrates. These genes are not expressed in reiterated domains early in embryogenesis but in a single spatially restricted domain in the developing brain (Fig. 7.11). Later in embryogenesis *En-1*, but not *En-2*, is expressed in a number of

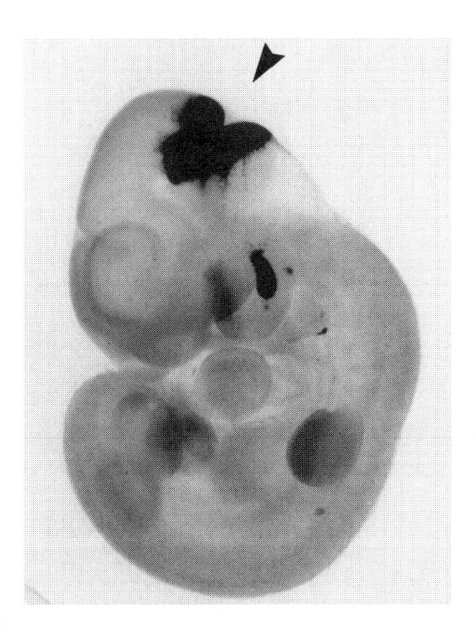

FIGURE 7.11 Expression of the *En-2* gene in a band across the mid/hind brain. The 10-day transgenic embryo contains a reporter construct with *En-2* DNA regulatory sequences expressing *lacZ*. The cells expressing *lacZ* (black under arrow) are visualized by histochemical staining. (*Photograph courtesy of Cairine Logan.*)

other domains in different tissues. Still later in development and in the adult, both genes are expressed in specific groups of neurons, but not in reiterated neurons as in the fly. From this descriptive analysis it was clear that the vertebrate *En* genes cannot control the development of reiterated posterior compartments. However, it was still highly likely that the genes act to direct the development of domains.

The most rigorous test of this hypothesis would be to make null mutations in *En-1* and *En-2*, study the phenotype of each single mutant, and then breed the mutants and study the phenotype of mice carrying null mutations in both *En* genes. This experiment would address the question of whether these structurally related genes carry out similar, or redundant, functions and thus represent backup systems.

The *En-2* gene was targeted in a male ES cell line using a vector, similar to the one shown in Figure 7.5, in which part of the gene was replaced with *neo* containing a promoter.[4] No enrichment scheme was used, since *En-2* is expressed at very low levels in ES cells and since the experiments were done before the PNS strategy was reported. Using the PCR strategy to screen for homologous recombinants, approximately 1 in 300 $G418^R$ clones was found to have undergone the predicted double crossover replacement event. Three targeted ES cell clones were isolated, of which only one gave rise to germline chimeras. Of the ES-derived offspring from the germline chimera, 50 percent were, as expected, heterozygous for the *En-2* mutation called *En-2^{hd}*. When mice heterozygous for the mutation were bred, they gave rise to offspring with the expected Mendelian ratio of 1 wild type : 2 heterozygotes : 1 homozygote *En-2^{hd}*. Furthermore, the *En-2^{hd}* homozygous animals had no detectable structural or behavioral phenotypic abnormalities. However, the one tissue, the adult cerebellum, that expresses *En-2* and not *En-1* was found by histological analysis to have an abnormal pattern of folds: Some folds were missing and others were misshapen.[5] Since the cerebellum controls many sensory–motor coordination functions, it is likely that these mice were deficient in some motor control functions but that the defects were not obvious in mice living in a cage supplied with amenities. Since the only obvious defect was in a tissue that expresses only *En-2* and not both *En* genes, the phenotype analysis suggested that *En-1* and *En-2* are functionally redundant.

Targeting experiments designed to mutate *En-1* in ES cells were not as readily successful as *En-2* experiments. A number of targeting vectors based on the PNS strategy, including different promoters to drive the *neo* gene and different regions of the *En-1* gene to direct homologous recombination, were tested in ES cells, and only one vector produced recombinants. The frequency of targeting *En-1* with this vector was 1/30 neo^R/GANC^S cells, whereas the total with the other vectors was less that 1/10^8 cells treated. The key change that was made with the successful vector was that cloned DNA from ES cells (**"isogenic"** DNA) was used for the arms of homology, rather than cloned DNA from mice of a different genotype. These results underscore the importance of near-perfect homology between the targeting vector and endogenous sequences for recombination to occur at a high frequency (see Chapter 3).

Future Prospects

At present, we have the very powerful capability of making specific mutations in mouse ES cells using gene targeting and transmitting the mutations into mice by ES cell chimera formation. However, a number of problems with gene targeting limit the kinds of mutations that can be made. One problem that has been discussed is that all genes may not be accessible to homologous recombination at detectable levels for two obvious reasons: (1) a selectable marker gene cannot be expressed after integration into certain genes in ES cells; (2) some sites in the genome may recombine at very low frequencies.

One possible avenue to explore in cases where the former is true is new targeting strategies in which the vector contains no selectable marker. In one such reported approach, the targeting vector was injected directly into the nuclei of ES cells instead of being introduced into the cells by electroporation.[6] Using microinjection of DNA into nuclei, approximately 10 percent of the surviving cells can integrate the DNA nonhomologously. This percentage is 100 times higher than in electroporation. Thus, if the ratio of homologous to nonhomologous recombination is 1 : 100, then by screening 1000 injected cell clones using the PCR strategy one targeted cell line could be identified. There are two drawbacks to this approach. One is that since ES cells are very small, microinjection into them is difficult. Therefore the injection of 1000 cells, or 3000 cells to obtain three targeted cell lines, could take one person weeks. The second drawback is that the frequency can be orders of magnitude lower than 1 in 100. Thus for some genes this approach is not applicable. A second possible technique for introducing DNA into cells, which has not been extensively explored, is to add to cells in culture short (less than 100 bases), single-stranded pieces of DNA that include the desired mutation. A possible advantage of this technique is that single-stranded DNA may be more recombinogenic than double-stranded DNA. For this approach to be useful, tens of thousands of colonies would likely have to be screened. This would require refining the standard sib selection procedure or mechanizing picking portions of colonies.

To tackle the potential problem of regions in the genome that undergo recombination at very low frequencies, further research into the mechanism of recombination will be needed. As more parts of the mammalian recombination machinery are characterized and DNA sequences that initiate recombination are identified, improvements to targeting approaches will likely become apparent. For example, in addition to using isogenic DNA, there is already evidence that specific DNA sequences introduced into a targeting vector can moderately stimulate homologous recombination.

A second limitation to the present targeting strategies is that they all include integration of a selectable gene into the targeted locus. For experiments designed to alter specific protein coding domains or to delete or add transcription regulatory elements, it would be ideal to make only the desired subtle alterations. On paper it is possible to design targeting vectors that can make subtle changes to the target gene by placing the selectable marker gene in a part of the endogenous locus where it will not interfere with the functioning of the gene. However, since the DNA regulatory elements that control transcription of a gene can be scattered throughout the gene and be many thousands of base pairs away from the gene, it can be problematic to identify a "neutral" site for insertion of the selectable gene. It would therefore be ideal not to have to insert a selectable marker into the targeted locus.

One area of research for which it is essential to make subtle mutations is structure/function analyses. A second area for which subtle mutations in mice will undoubtedly play an important role is medical research. One active area of medical research is the study of animal models of human diseases. For some gene mutations that affect humans as diseases, mice are available that contain mutations in the same gene. In some instances the disease symptoms of the mice are very similar to those of humans. These animals serve as models of the human diseases and are used for further research into the biochemical bases of the diseases. Potentially they can also be used to test new therapies for the diseases.

Some mouse mutants do not, however, have the same set of disease symptoms as humans with mutations in the same genes. In some cases this is because the physiology of mice and humans is not exactly the same. This fact can nevertheless give insights into the biochemical processes that involve the affected genes. In some cases, the differences between the mouse and human symptoms may be due to the types of mutations in the two species. One mutation may be a null mutation whereas the other may be a leaky mutation. The most direct way to test which is the case would be to use germline gene targeting to make mice with exactly the same mutation as the humans. In this type of gene targeting, of course, a marker gene cannot be included. Finally, for the majority of human diseases animal models do not exist, and germline gene targeting represents the most direct way to make such models.

The pop-in/pop-out and direct gene replacement procedures described for yeast in Chapter 5 represent two-step schemes for making subtle mutations that could be applicable to ES cells (Figs. 7.12, 7.13). There have been two recent reports of successful targeting using the pop-in/pop-out approach in ES cells. In one case, the single hprt gene on the X chromosome of a male ES cell line was mutated.[7] Since Hprt-expressing and Hprt-nonexpressing cells can be directly selected, this gene served during the homologous recombination step as both the positive and negative marker for the two selection steps. In the second case, an autosomal gene was mutated at a surprisingly high frequency, with the pop-out or intrachromosomal recombination occurring at a frequency of 1 in 1000 cells.[8]

As in yeast, the potential limiting factor for the pop-in/pop-out and direct gene replacement approaches in ES cells is the frequency at which the negatively selectable marker gene spontaneously becomes nonfunctional. The marker gene can become inactivated by at least three mechanisms. One is suffering a spontaneous mutation, such as a deletion, insertion, or base pair change. This event usually occurs at a low frequency (1 in 10^6 cells/cell doubling) in somatic cells and should therefore not represent a major limiting factor for these approaches. A second

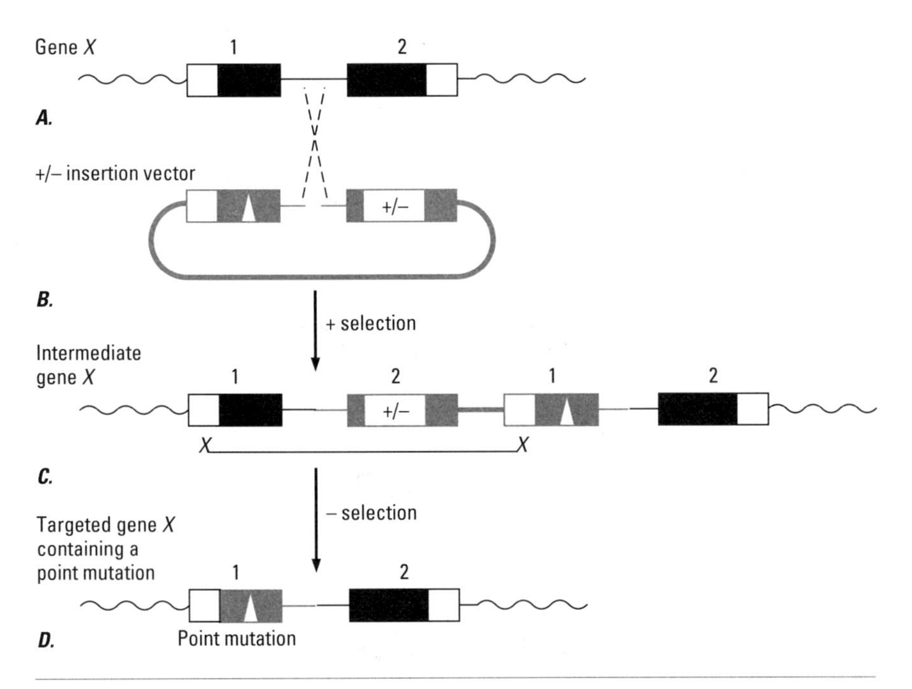

FIGURE 7.12 Possible scheme for producing a point mutation in ES cells using the pop-in/pop-out targeting approach. (*A*) Gene *X* shown as in Figure 7.5*A*. (*B*) Insertion targeting vector cut in the region of gene *X* homology. The vector contains a point mutation (white arrowhead) in the first exon and a marker gene (+/–) that can be selected both for and against. The vector is electroporated into ES cells and the cells are grown in medium that selects (+) for expression of the marker gene. The colonies surviving selection are screened for targeted clones. (*C*) Structure of gene *X* in the intermediate cell line following homologous recombination and insertion of the targeting vector. The intermediate cell line containing the targeted gene *X* are grown in medium that selects against (–) cells expressing the marker gene. Some of the cells that survive negative selection will have undergone an intrachromosomal recombination event between the inserted gene *X* sequences and the endogenous sequences. (*D*) If the recombination occurs upstream of the point mutation (indicated by X's) then the point mutation will be retained.

mechanism for inactivating genes, which does not alter the gene sequences themselves, is DNA methylation. Somatic cells actively methylate C residues in DNA, and it has been found that most highly methylated genes are not transcribed. The frequency with which a marker gene is inactivated by this process will probably vary from site of integration to site of integration and from cell line to cell line. The third mechanism for inactivating a gene is gene loss. It is not known what is the most prevalent mechanism of gene loss in ES cells, but it is becoming clear that marker genes can be lost at a high frequency (up to $1/10^3$ cells/cell generation) in

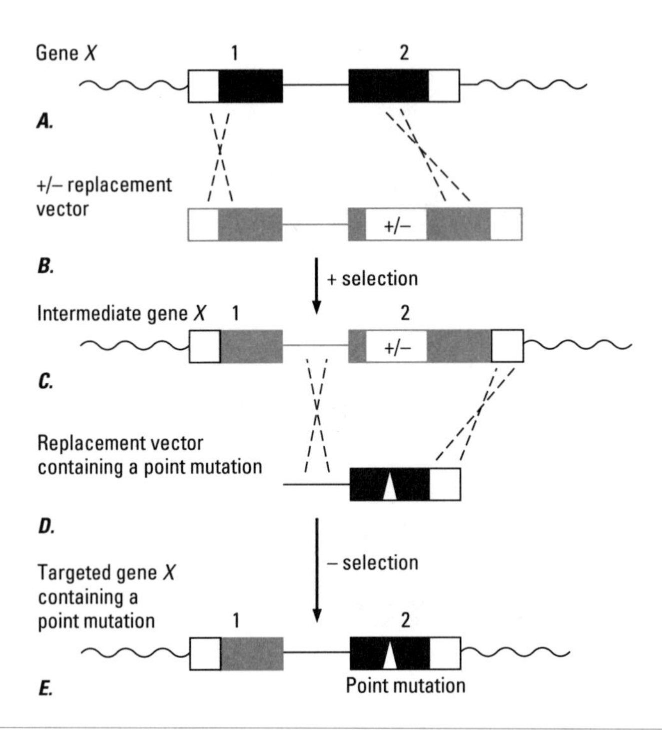

FIGURE 7.13 Possible scheme for producing a point mutation in ES cells using the direct gene replacement targeting approach. (*A*) Gene *X* shown as in Figure 7.5*A*. (*B*) First replacement vector, similar to that shown in Figure 7.5*B* except that the marker gene can be selected both for and against. The vector is electroporated into ES cells, the cells grown in selective medium, and targeted clones identified as described in the legend for Figure 7.12*B*. (*C*) Structure of gene *X* in the intermediate cell line following homologous recombination with the replacement vector. (*D*) Second replacement vector, containing gene *X* sequences with a point mutation. The second vector is electroporated into the intermediate cell line, and the cells are grown in medium that selects against expression of the marker gene. Homologous recombination between the second replacement vector and the targeted gene X will result in replacement of the marker gene with the point mutation. Cells containing the point mutation will survive back selection.

ES cells. If all genes can be lost at this high frequency, the pop-in/pop-out and direct gene replacement approaches in their present form will be limited to genes for which the second recombination event occurs in greater than 1 in 10^6 cells.

At present, therefore, there is no straightforward method for making subtle mutations in most genes in ES cells. Improvement of the existing techniques and design of new approaches is certain to represent an important part of future research into gene targeting technology. In addition, further research into the growth requirements of ES cells should improve the efficiency of producing germline chimeras from targeted ES cell clones. Finally, the main thrust of germline gene targeting experiments probably will be to accumulate new mouse mutants and analyze their mutant phenotypes in order to gain insights into the biological functions of the mutated genes.

Notes

1. Ecker, R. (1929). The recessive mutant *engrailed* and *Drosophila melanogaster. Hereditas* 12: 217–222.

2. Poole, S. J., L. M. Kauvar, B. Drees, and T. Kornberg (1985). The *engrailed* locus of *Drosophila*: Structural analysis of an embryonic transcript. *Cell* 40: 37–43.

3. Joyner, A., T. Kornberg, K. G. Coleman, D. Cox, and G. R. Martin (1985). Expression during embryogenesis of a mouse gene with sequence homology to the Drosophila *engrailed* gene. *Cell* 43: 29–37.

Joyner, A. L. and G. R. Martin (1987). *En-1* and *En-2,* two mouse genes with sequence homology to the *Drosophila engrailed* gene: Expression during embryogenesis. *Genes and Development* 1: 29–38.

4. Joyner, A. L., W. C. Skarnes, and J. Rossant (1989). Production of a mutation in the mouse *En-2* gene by homologous recombination in embryonic stem cells. *Nature* 338: 153–156.

5. Joyner, A. L., K. Herrup, A. Auerbach, C. A. Davis, and J. Rossant (1991). Subtle cerebellar phenotype in mice homozygous for a targeted deletion of the *En-2* homeobox. *Science* 251: 1239–1243.

6. Zimmer, A., and P. Gruss (1989). Production of chimaeric mice containing embryonic stem (ES) cells carrying a homeobox *Hox* 1.1 allele mutated by homologous recombination. *Nature* 338: 150–153.

7. Valancius, V., and O. Smithies (1991). Testing an "in-out" targeting procedure for making subtle genomic modifications in mouse embryonic stem cells. *Mol. Cell. Bio.* 11: 1402–1408.

8. Hasty, P., R. Ramirez-Solis, R. Krumlauf, and A. Bradley (1991). Introduction of a subtle mutation into the *Hox* 2.6 locus in embryonic stem cells. *Nature* 351: 234–246.

Suggested Readings

Capecchi, M. (1989). Altering the genome by homologous recombination. *Science* 244: 1288–1292.

Joyner, A. L. (1991). Gene targeting and gene trap screens using embryonic stem cells: New approaches to mammalian development. *BioEssays,* 13, 649–656.

Joyner, A. L. (1993). *Gene targeting: A practical approach.* Oxford: IRL Press, in press.

Joyner, A. L., and Hanks, M. (1992). The *engrailed* genes: Evolution of function. *Seminars in Developmental Biology* 13: 1–8.

Rossant, J., and Joyner, A. L. (1989). Towards a molecular genetic analysis of mammalian development. *Trends Gene.* 5: 277–283.

EPILOGUE

At present, gene targeting has important applications in basic genetic research. Is there a potential for direct applications in human medicine? As with many discoveries made in basic research, the answer is yes. Such applications would come under the general heading of gene therapy, which is, quite simply, medical therapy achieved by changing the genetic information encoded in the human genome. The very notion of gene therapy is charged with controversy and has been dramatized by the popular press. Not without reason, since certain aspects of gene therapy raise important ethical considerations. The procedures that make gene therapy possible have been developed in mice, and some are already being applied to humans. Thus the time is here for everyone, not just scientists, to thoroughly explore the issues surrounding human gene therapy.

In the following discussion we present some of the facts and issues surrounding gene therapy in order to provide a framework for further thought and discussion on the subject. We have intentionally refrained from expressing our personal views on ethical issues. We believe that everyone should have the opportunity to contribute to the discussion on how far human gene therapy should be taken.

What Is Gene Therapy?

Gene therapy is a broad term that encompasses all methods of altering human genetic information. It can be subdivided into a number of distinct categories. The first and more profound distinction is whether genetic information is being changed in somatic or germ cells. Somatic gene therapy is the genetic engineering of a specific subset of the cells in the body, for example, blood cells or liver cells. Germline gene therapy, on the other hand, involves changing the genetic information in germ cells or fertilized eggs. As with transgenic mice (Chapter 6), all the cells of a human being that develop from a genetically manipulated egg would contain the introduced genetic alteration. The ethical issue raised is that the introduced genetic alteration could be propagated in some of the offspring of that individual, and thus could spread through future generations.

The second, more practical distinction is whether genetic information is being changed by homologous or nonhomologous recombination. Gene targeting obviously lies in the first category. Since mammalian cells normally undergo nonhomologous recombination at a much higher frequency than homologous recombination (Chapter 5), the practical implication is that gene targeting is currently feasible only in systems where large numbers of cells can be obtained, treated with DNA, and screened for rare homologous recombination events.

Somatic Gene Therapy

Somatic gene therapy of humans can be achieved, in a limited way, using extant techniques. In general terms, human cells are removed from the body and placed into culture, DNA is introduced, and the cells are returned to the body. To make the process effective, either the majority of the cells must incorporate the DNA in the desired fashion, or some method must be used to isolate the cells with the desired genetic change. Given the techniques available, this presents a severe limitation for approaches that require targeted homologous recombination rather than random integration. Two possible advances, development of which certainly is not beyond the realm of feasibility, could overcome this limitation. First, it may be possible to develop methods that would greatly increase the frequency of homologous recombination. For example, an appropriate recombinase enzyme may be discovered, purified, and coinjected along with the DNA. Second, improved tissue culture methods may be developed that would allow the propagation of primary human cells in culture for extended periods of time, so that rare targeted cells could be isolated and expanded before being returned to the body.

The limited proliferative potential of many human cell types when explanted directly from the body is a serious limitation for somatic gene therapy using strategies of either nonhomologous or homologous recombination. Considerable research efforts have been and continue to be directed at this issue. In addition, in most circumstances it will be necessary to identify, selectively grow, and genetically manipulate pluripotent stem cells that can effectively reconstitute a tissue when grafted back into the body. For example, truly pluripotent stem cells of the immune system have not yet been isolated, despite intensive effort over many years. Nevertheless, research in these areas is progressing rapidly, and major developments with applications for somatic gene therapy are likely in the near future.

Methods also are being developed to introduce DNA directly into cells resident in the body. This approach obviates the need to explant cells and propagate them in culture. One such technology is the DNA particle gun. Tiny projectiles are coated with DNA and then accelerated at high velocities toward the exposed tissue. The principle is that of a shotgun, except on a very small scale. Some of the projectiles enter individual cells without killing them. Once in the cytoplasm, the DNA dissolves from the surface of the projectiles and becomes available for recombination in the nucleus. The major disadvantage of this method is that the yield of cells that have stably incorporated the DNA is currently below that achievable with cells in culture.

Somatic gene therapy for the cure of a heritable metabolic disorder, adenine deaminase deficiency, has already been approved by the regulatory agencies and is in clinical trials. More applications can be envisioned, and a number are in the design stage. The list of heritable disorders with a metabolic basis that might be tackled using gene therapy goes on and on. Examples include sickle cell anemia, diabetes, cystic fibrosis, and muscular dystrophy. Not surprisingly, this is an area of intensive research.

Germline Gene Therapy

The production of transgenic humans by the route of zygote injection is, with current technology, theoretically feasible. The efficiency, however, of producing transgenic individuals would be so low that it would not be practically feasible. Extrapolating from research on the production of transgenic mice, approximately 100 fertilized eggs would have to be injected to produce one transgenic human. Effective production of transgenic humans would therefore require developing new techniques that would dramatically increase the efficiency. Application of germline gene targeting by zygote injection would require additional technological breakthroughs to increase the frequency of homologous recombination to near 100 percent. Although such breakthroughs are not likely to come very soon, they are not beyond the realm of possibility.

Germline gene targeting using the techniques that have been developed in mice would not be possible in humans unless human ES cells became available. The isolation of human ES cell lines poses ethical issues outside the realm of gene therapy. The most profound point is that isolation of human ES cell lines would open the way to producing clones of people identical to the embryo from which the cells were derived. Furthermore, additional ethical questions are raised by the idea of making human chimeras. What would be the source of the host blastocyst? If the host blastocyst is taken from different parents than the ES cells, then what are the rights of each set of parents to the chimeric child that is born? There is no obvious scientific reason why human ES cells could not be established, and the techniques of in vitro fertilization have opened up the possibility of making chimeras. It is therefore essential that a decision be made soon on whether human ES cells and human chimeras are ethically acceptable.

Is Gene Therapy Ethically Acceptable?

Three issues must be considered when any new medical intervention is contemplated. First, the therapy must result in a substantial improvement in the quality of life for the patient. The second point is the availability of other means of therapy. For example, it is debatable whether it is desirable (and cost effective) to use gene therapy for diseases for which there are other widely accepted and generally adequate treatments. The third issue is the potential for unforseen side effects along the way.

Unquestionably, the patient is likely to be concerned with the third point. It is widely accepted, however, that essentially all medical procedures carry some finite risk of serious, unintended consequences. For example, approximately one in every 300,000 children who receive a pertussis vaccination will suffer permanent brain damage from the vaccine. Approximately one in every 10,000 patients subjected to general anesthesia dies from complications resulting from the procedure. Nevertheless, the majority of our children are vaccinated and people routinely undergo general anesthesia for elective surgery. Laying aside the economic issue of cost, the choice becomes one of benefit versus risk. Most medical procedures necessitate such choices, and we are accustomed to making them. We have been making such choices from the very first days of medicine.

Is gene therapy any different? Can we go about it the same way we go about other medical procedures? It could be argued that somatic gene therapy falls into this mold. First, lethal diseases exist for which there are no cures. If successful, in these cases gene therapy may offer the only hope. Second, if anything unexpected happens, only the patient suffers the consequences. When that individual dies, the genetic mishap would therefore be lost.

Germline therapy, however, requires new arguments, since any genetic mishaps can be propagated to future generations. In order to evaluate the question of whether germline therapy should be allowed, we will consider the possible benefits and risks. Since ES cells carry with them many additional ethical questions outside gene therapy itself, we will continue the discussion with the assumption that zygotes will be used. Furthermore, we will also assume an ideal situation where DNA can be inserted by nonhomologous as well as homologous recombination at a high frequency using techniques that are economically feasible. What benefits could the human race gain from such a procedure?

In terms of curing genetic diseases, germline therapy does not seem to offer a considerable benefit. Such therapy, of course, would be of no value to living individuals who are suffering from a disease, since it is only their offspring who could be cured. Furthermore, offspring from an individual homozygous for a recessive genetic disease would automatically be cured by the normal gene contributed by the second parent. In cases where a disease is dominant, or two heterozygous individuals wish to have children, an alternative is already widely used. Medical research has provided methods for prescreening embryos for genetic defects and the parents of affected fetuses can make the decision whether the child should be born.

If germline therapy is not worthwhile for curing heritable defects, then the question becomes whether any other potential benefits can be found. Could we add new genes to our genomes (or take potential troublemakers away) that would make our future generations "superhuman"? If we can decipher the genetic programs that make us what we are, can we also reprogram them, for example, to extend our natural lifespan or make us less susceptible to cancer? A seductive argument can be made that there is no fundamental biochemical reason that we should be locked into a given lifespan or be ultimately consumed by malignancies. After all, mice live for three years, humans for about seventy, and some turtles for up to three hundred. Why? The underlying reasons behind cancer and aging are likely in our genes, and thus they are not an absolute given, but rather are changeable.

What about the risks of germline gene therapy? The main point is, Can we predict the outcomes and side effects of reprogramming human beings? The answer, obviously, is no. The fundamental distinction of germline therapy is that the genetic alterations, deliberate or inadvertent, have the potential of being stably inherited and thus spreading through human populations. Is it worth tinkering with evolution by changing our genetic makeup if we cannot predict all the outcomes? How could we accommodate the issue of accountability for what are essentially long-term experiments on humans, since the possible effects of such therapy are not limited to the individual but extend to the society, the environment, and even the future of the earth as we know it today? Is this a fundamental difference from existing medical applications or is it simply an extension of the age-old benefit versus risk assessment?

The issues are complicated and there are no simple right or wrong answers. Where do we draw the line? Should it be at basic research, somatic gene therapy, germline gene therapy, or not at all? On the one hand it can be argued that a clear distinction should be made between basic and applied research. Basic research involving genetic engineering is providing a framework as well as direct clues for understanding the mechanisms responsible for human diseases, whereas direct applications raise many ethical questions that are not easily answered. An alternative argument is that basic research itself raises ethical questions by the very fact that it could be applied in a way that might create unforseen harm. Such discussions are not specific to gene technology but are relevant to all aspects of basic research. This is not the first time that ethical questions of this kind have surfaced. What is not yet in place for gene therapy but should be developed is an effective forum for discussing the issues, coming to conclusions, and ultimately enforcing the decisions. As scientists it is our responsibility to inform the public from an objective viewpoint of the possible outcomes of using (and misusing) available technology. However, the long-term implications raise ethical issues that must be discussed by all of us.

INDEX

Note: Page numbers in *italics* indicate illustrations; those followed by t indicate tables.